# My Shadow

by

Robert Brun

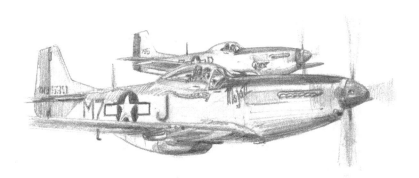

# Robert Brun

**Aire Apparent Publications**
76 State Street
Newburyport, MA 01950

Copyright © 2013 by Robert Brun

All rights reserved. No part of this book may be reproduced, stored or transmitted in any form or by any electronic or mechanical means without written permission from the publisher, except by a reviewer who may quote brief passages in a review.

Printed in the United States of America

ISBN 978-1-931376-40-2
10 9 8 7 6 5 4 3 2 1

# My Shadow

## Dedicated to

**Francis (Frank) Bloomfield Brun Sr. 1912-1999**

To my father
2nd Lt. Francis Bloomfield Brun Sr., 5th Army, O.S.S.
And
To all the brave men and women who
Served their country in its time of need.

**Robert Brun**

# My Shadow

## Contents

| Chapter | | Page |
|---|---|---|
| Prologue (Present Day) | | 9 |
| 1 | Waiting | 13 |
| 2. | Abington | 28 |
| 3. | Training | 36 |
| 4. | Baptism | 44 |
| 5. | My Shadow | 51 |
| 6. | Over France (Pt 1) | 57 |
| 7. | Over France (Pt 2) | 64 |
| 8. | Over France (Pt 3) | 68 |
| 9. | First Five | 73 |
| 10. | Back From the Pub | 81 |
| 11. | Big Week | 86 |
| **Part II** | | |
| 12. | Scrubbed | 97 |
| 13. | Ace + | 101 |
| 14. | Transitions | 112 |
| 15. | London Leave (Pt 1) | 119 |
| 16. | London Leave (Pt 2) | 132 |
| 17. | Day of Days | 135 |
| 18. | A New Ride | 144 |
| 19. | Letter From Home | 150 |
| 20. | Ground Attack | 155 |
| **Part III** | | |
| 21. | Transfer | 161 |
| 22. | Tiger, Tiger Burning Bright | 167 |
| 23. | R & R | 173 |
| 24. | Thanksgiving | 180 |
| 25. | The Bulge | 183 |
| 26. | Base Attack | 188 |
| 27. | G-Suits | 195 |
| **Part IV** | | |
| 28. | Trains | 203 |
| 29. | St. Valentine's Day Massacre | 209 |
| 30. | A New Threat | 213 |
| 31. | Flight Home | 223 |
| 32. | Arrival | 234 |
| 33. | Boys to Men | 238 |
| 34. | All Right | 243 |
| 35. | The Shadow | 247 |
| Epilogue | (Present Day) | 249 |
| | Glossary | 251 |
| | Aircraft Types | 253 |
| | About the Author | 255 |

**Robert Brun**

# Acknowledgment

I would like to thank the following people without whom this book would not have been possible.

**Captain Joseph M. Dwyer** (USAAF) who, unknowingly, gave me the idea for this book.
**Nancy Alcorn** who patiently waited for and edited the bits and pieces of manuscript as they came in.
**Joe Buschini** who showed me how to write a better sentence.
**2nd Lt. Francis *"Frank"* Bloomfield Brun Sr.** for the stories he shared with me about his service with the 5th Army and the O.S.S. during World War II. I wish we'd had more time together.
    And finally, all the brave men and women whom I have met and have shared personal stories with me of their service during that horrific war. May it never again become necessary to do so.

Robert Brun

# My Shadow

## Forward

People have asked me why I am so interested in the Second World War. Well the simple answer is because my Father was part of it. It was, however, not until a few years before his death that I came to learn what part he'd played. Like tens of thousands other veterans who returned from this global conflict, he did not speak much of the war and when I finally asked why, his response surprised me. *"I didn't think anyone was interested."*

Unfortunately, until in my late twenties, this statement was, more or less true. Although I had constructed plastic models of virtually every plane from that era, my interests had always been with the aesthetic and not the historic aspects of that time. Since then, I have come to understand how this period, more than any other shaped the world we live in. My Father lived through it, participated in it, and brought back his own stories.

He is no longer with us, but since his passing, I have met a great many veterans from that era. Congressional Medal of Honor recipients and Grunts, Flying Tigers and Tuskegee Airmen, WASPS and Med Station Nurses, and each and every one have been the definition of Ladies and Gentleman. I have come to respect these people and what they gave up in order to stop the spread of Fascism, Nazism, and Imperialism. These young men and women, who grew up during the "Great Depression," were called upon to serve. They did their part came home and got on with their lives, many did not. They knew sacrifice and appreciated that some things are worth fighting for. It is through their efforts that we have been able to enjoy the freedom we grew up with.

During my travels as an Aviation Artist throughout the country, I have come to see that far too many younger people have no understanding of this period of world history other than the "*awesome planes*" it produced. This has always troubled me and so far, it has been through my paintings and lectures that I make the effort to educate the viewer. Now I bring this desire to the written word.

Throughout the writing of <u>My Shadow</u> I have made a concerted effort to be as accurate as possible. The dates and the events, which take place within the story are real and did occur, but of course from a fictionalized point of view. I have also included a glossary of terms and phases from that period that may be foreign to today's readers. In writing this book, I have endeavored to educate as well as entertain and perhaps give the reader the desire to learn more about this remarkable period of world history.

War is never a good idea and should always be seen as a last resort, but sometimes it becomes necessary. In 1941, for the people of the United States, it became necessary and this is my tribute to those who did what was required. Thank you.

R. Brun 2013

**Robert Brun**

**My Shadow**

# Part I
**Prologue**
Present Day

Joe pulled the car into the driveway and switched off the ignition. He sat there listening to the slow tick of the cooling engine, not sure what to do next. He had never enjoyed these mandatory insurance physicals, but up until now it had always been the same,
"*Turn your head and cough... Come back next year.*"
"No wonder health insurance is so expensive." he thought, "What a waste of time." Today, however, the doctor's report had not been routine.
Joe wasn't really surprised. His health couldn't last forever and had been ignoring the mild chest pains for some time now. Indigestion, he'd told himself, the Doc would merely suggest that he change his diet and cut back on his golf game. So the recommendation for a battery of tests and possible heart surgery had come as a shock.
Joe reached into the pocket of his sport coat for the pack of cigarettes that wasn't there and hadn't been since the evening his wife had threatened to leave him if he didn't quit. He knew that she wasn't serious about the leaving part, but the look on her face told him there would be no further discussion. Joe knew that *look* and quit smoking the very next day.
For more than twelve years Joe had kept his word, though he still got the craving for a smoke whenever stress got above a certain level, and

this was certainly one of those times. The idea of open-heart surgery did not appeal to him in the least.

Joe unfastened the seatbelt and grimaced as he lowered first one, then the other leg out of the car and onto the driveway. Getting old was certainly not for the faint of heart and now his own heart, like his knees, was giving out too. The doctor was right though; the tests would show what Joe already suspected, he had all the symptoms of a blockage and there was no use kidding himself about it any longer.

Closing the door, Joe depressed the key fob and listened for the familiar *chirp* that locked the vehicle. He looked up at the house he and his wife had shared for almost sixty years and wondered how he was going to tell her this choice piece of news. In their sixty-four years of marriage, they'd been through it all, good times and bad, but he hadn't had to fear for his life since his discharge from the Air Force or the *Army Air Corps* as it had been called in 1945.

The house was small by today's standards and positively dwarfed by the rambling estates that filled the numerous sub-divisions that had sprung up around them over the last six decades, but Joe didn't care; he loved the little house. It was this house they had purchased back in 1951 with the down payment money he and his wife had saved during those first five lean years after the war.

The G.I. Bill had made it possible for Joe to pursue his college degree in business management and get a good job, something he'd never dreamed of coming from the dairy farm where he'd grown up. It had been hard, but it had been worth it. Well into his *golden years*, with two grown kids and three grandchildren, life had been about as good as he had dared hope. Now this.

Climbing the steps to the porch, Joe checked the mail and riffling through the bills, letters and various catalogs, almost tripping over a cardboard box sitting by the door. Bending down to inspect the parcel, Joe read the address:

Captain Joseph M. Dyer
1642 Carmichael Way
Fayetteville, GA 32186

Captain? Joe wondered. He hadn't been addressed by his rank since his military discharged over sixty years ago.

The box lacked any kind of shipping label, just the address in black marker written in a shaky hand. Joe bent to lift the package and was surprised by its weight. After opening the door and with a grunt, Joe slid the heavy box into the house.

## My Shadow

      No one was home. His wife was spending the afternoon with their youngest grandson and wouldn't be back until that evening so after sorting the mail and checking the answering machine, two sales calls and a message from the lawn guy, Joe hefted the box up onto the kitchen counter.
      Using the blade of the penknife he kept on his keys, Joe slit the tape and pushed opened the cardboard flaps trying not to spill the styrofoam peanuts that filled the box. He opened the plastic bag within and was struck by a strong odor of old paper and leather, like the musty smell of an antiquarian bookstore.
      Confused, but at the same time curious, he lifted first one, then two, then a stack of worn leather bound books from the box. Each volume had a large white hand painted number on the cover with a date beneath it. The first read.

*Volume #1 597th FS 11/19 - 12/28/43*

      Joe could feel his pulse pounding in his temples and picked up a padded manila envelope that accompanied the books pulling out a small, square, metallic object covered in bubble wrap.
      Carefully peeling off the tape, he held up a plastic bag containing a Zippo lighter with its familiar smell, and a note with just three words:
      *"The Shadow Knows!"*
      Joe stared at the letters M.C.C. engraved into its scratched cover and sat down hard on the barstool.
      *"Shadow."* The name flashed through Joe's mind. An uneasy feeling accompanied the recollection.
      Staring at the pile of musty books, Joe now understood, these were the journals that contained the events of so many years ago. A record of the time when no one had a choice about what needed to be done. A time when so many didn't come back, and the ones that did survive just wanted to put it all behind them and forget, to get on with their lives. This is exactly what Joe had done. Now he was being asked to remember those painful times again.
      He resented the intrusion of this past into his present life, and at the same time felt ashamed. He dreaded the thought of reviving the memories, but because of what they had been through together and what had happened, he owed *him* at least that much.
      Joe got up and made a strong pot of coffee trying hard to relax. It was, after all, over sixty-five years ago and so much had changed since then, at least Joe thought it had. He could do this; he would remember again he would do it for *the Shadow*.

## Robert Brun

Taking his mug down from the cabinet, he poured himself a coffee, added the milk and two sugars, and sat down at the counter. With a trembling hand he reached for the first book. Carefully opening *"Volume One,"* Joe's eyes immediately fell upon a sketch of a thin, worried looking Lieutenant sitting on the deck of a troop ship, holding a cup of coffee. The drawing captured both the excitement and anxiety of this young man.

The words *"Lt. Joe"* were scribbled beneath the drawing in a distinctive hand and Joe flushed. He reached into his breast pocket for a cigarette while at the same time expertly flicked open the lighter and thumbed its wheel.

The Zippo lit, but with no cigarette to smoke, Joe just sat there staring into the orange flame. It was 1943 all over again…

# My Shadow

## Chapter 1
## Waiting

*(Journal entry)*
*Abington, England Friday, April 6, 1945: An awful day, particularly after the night I had. I'm still not certain what happened, but I'm not in the mood to find out. Maybe after today's mission, maybe - R.B.*

The last light of the day was fading from the western sky over East Anglia, the eastern edge of Great Britain. Glancing at his watch, Joe noted the time. 22:35, 10:35 p.m. civilian time, a late sunset for someone from Georgia, but double-daylight savings time was only one of many adjustments to his life since shipping out of New York City that cold, wintry day sixteen months earlier.

Captain Joseph M. Dyer stood staring while the sun disappeared below the horizon. Straining his eyes, he scanned the sky for the silhouette of the last of the 597th Fighter Group's P-51s due to return from the day's mission. His excellent eyesight that served him so well in combat was of little use now as even the tree line grew indistinguishable from the darkening sky.

How many missions had there been? Joe had lost track, but for the last year his wingman had always been with him, "Covering your ass!" as Rob was so fond of saying, while letting Joe take all the credit.

"You do what you want with your plane" Rob would joke, "but I don't want any of those stinkin' Nazi swastikas marking up my plane, not on my *Shadow.*"

*Shadow.* That's what Rob called his plane and that's how Joe thought of his wingman. *Shadow,* the name seemed to suit him. No matter what Joe did, how he turned or where he flew, somehow Rob was always there, watching his six o'clock and keeping Joe's tail clear of the *Bratwursts* or just *Brats* as Rob called the German fighters.

"*Krauts* is too good a name for them," Rob would say joking, "and besides, I like Kraut, can't stand Bratwurst. Awful stuff."

Glancing over at the 12 swastikas his crew chief Zeke had stenciled along the fuselage, Joe knew the credit for most of those victories

## Robert Brun

were as much Rob's as they had been his to claim. Anytime he would try to make that point, however, Rob was quick to reply,

"You just concentrate on putting holes in those *Brat* fighters and leave clearing your tail to me."

And clearing it he did in too many missions to count, Joe had not one bullet hole to show for his time in combat.

Others had not been so lucky, and luck, Joe knew, attributed for most of it. After all the training and the experience, much of what got any pilot through was nothing more than Lady Luck, and if the *Lady* wasn't smiling on you, watch out! Many of the pilots he'd shipped in with had not been so fortunate.

For Rob, however, it was a different story. His plane had caught more rounds than you could count. The bullet and canon holes Rob brought back from so many missions had earned him the moniker *Lieutenant Swiss Cheese* from the guys over in maintenance. Joe wondered how many of those rounds had been meant for him.

It was Joe's opinion that Lieutenant Robert Browning was the best wingman in the 597th. Maybe even the entire 8th Air Force, although Joe was far from unbiased, but Browning would never have accepted such praise. Even before flying their first mission together Rob had demonstrated his uncanny ability to follow his Flight Leader through any maneuver. "Sometimes even on the ground" Joe thought, and he smiled at his recollection of their first mission.

Fighter pilots always flew in pairs, a Flight Leader and a Wingman with two pairs making up a Flight, four Flights a Squadron of sixteen planes and three Squadrons, a Fighter Group, forty-eight planes in all. Although pretty much on your own once the shooting started, each Wingman's job was to look out for his Flight Leader while the Flight Leader did his job, and that job was shooting down enemy planes.

A good Wingman was a pilot you could count on, one who would be there when you needed him; someone who would look out for you even when you didn't think you needed looking after and at the same time, do so without getting in your way. That's the kind of Wingman, you wanted and that was the kind of Wingman, and *friend* Lieutenant Robert L. Browning had been to Captain Joseph M. Dyer. Joe felt sick with worry.

### Sunday, November 28, 1943

Joe first met Rob during their Atlantic crossing from New York City. He, like hundreds of other newly commissioned pilots had just finished training and was finally being shipped out to the European Theater of Operations (E.T.O.) and a chance, he'd hoped, to actually make a difference.

## My Shadow

Since General Henry *Hap* Arnold had assigned Brigadier General Ira C. Eaker to establish the Eighth Air Force in England in the summer of 1942, B-17 *Flying Fortresses* and B-24 *Liberators* had been making tactical bombing raids into German-occupied Europe. Unlike the British, who bombed mostly at night under the cover of darkness, the USAAF had maintained that accurate, daylight bombing was the answer. Knock out the German's industrial ability to make war, and they'd soon surrender. That, at least, was the belief.

In the beginning, the USAAF also believed that these heavily armed *Fortresses* and *Liberators* could defend themselves against German fighters. This sadly, proved not to be the case. Although armed with up to thirteen 50-caliber machine guns each, the USAAF bombers forces were being decimated by the nimble and highly experienced *Luftwaffe* pilots. Continued losses like those suffered in early 1943 could not be sustained and as a result, Joe and several thousand pilots just like him were being trained for a new type of flying, that of the escort fighter.

Boarding the converted ocean liner on that cold, dreary afternoon from pier #3 in New York, Joe took his place in line with hundreds of replacement troops on their way to England. Like a gigantic olive-drab centipede, GIs made their way up the gangway and aboard the transport ship that had taken two full days to load. Approaching the sergeant at the top of the ramp, Joe called out his name and received his cabin assignment. Following the flow of troops, he descended a steep ladder into the interior of the ship.

The cold, damp air of the November afternoon was quickly replaced by the stale musty smell generated by the multitude of bodies that filled the ship. Managing to reach the correct deck, Joe made his way along the grey corridor until he found his assigned quarters.

Once an opulent stateroom, its luxurious accommodations had since been removed leaving nothing more than an eight-by-fifteen foot steel box painted with thick battleship grey enamel, the odor of which still hung heavy in the air. At one end of the cabin a single porthole provided a blast of cold air, but was insufficient to purge the stuffiness from the cabin. Lining both sides of the room were pipe berths stacked three high leaving little more than enough room to squeeze between them. A hanging closet and a small sink were to be shared by the twelve officers assigned to this cabin.

As the first to arrive, Joe stowed his gear in the closet and stretched out on second berth only to discover that his feet extended beyond its length. Feeling more like a sardine than a fighter pilot, he removed the letter from his girl and reread it for the tenth time. He wished it said more, but understood her concern about the uncertainty of their...

## Robert Brun

his future.  Sitting up, Joe banged his head on the bunk above and resolved to spend no more time in these cramped quarters then was necessary.  He left the cabin eager to explore the ship.

Walking back down the corridor, Joe happened upon an open space resembling a large storage area lined both sides and down two rows in the middle with the same pipe berths as in his own quarters, but these were stacked five high.  Hundreds of Army *grunts* literally crawled over one another to find their spot and the idea of sharing the crowded stateroom with eleven companions no longer seemed quite so bad.

Fighting the flow of boarding troops, Joe made his way back up on deck and into the fresh air.  Standing at the rail, he was surprised to see the pier crowded with people cheering and waving.  Many held signs that wished them well and personal, rude messages directed at *der Führer*.  Pretty girls waved handkerchiefs and blew kisses and small boys waved flags and saluted.  It was all very festive, but Joe only felt anxious as he watched the dockhands prepared the ship for departure.  They were finally on their way, and the excitement was palpable.

The ship eased out from the pier and was pulled by a flotilla of tugs past Liberty Island, where even the statue of *Lady Liberty* herself seemed to bid them farewell.  Despite the revelry both ashore and on board, Joe wondered if he would ever see any of this again.

Steaming out of New York Harbor, the transports joined up with the destroyer escort all heading east, and after several hours, the sun low on the western horizon Joe began to settle in for the long ocean voyage.

*"Now hear this...now hear this.  Mess will begin at 16:00 for ship's personnel and crew.  All other military personnel will eat at 18:00.  That is all."*

The message blared out over the ship's public address system.  Chow?  Joe hadn't given food much thought with all that was happening, but now realized he was hungry.

He didn't have the first clue where the mess was and by the time he had located it a line of soldiers waiting to be fed stretched from the galley hatch halfway around to the fantail.  Meals, Joe learned, were one place where his Lieutenant's rank made little difference and forty-five minutes later, Joe entered the large open room where the troops were fed.

Men stood at stainless steel tables, wolfing down their food.  The heat from the kitchen mixed with the humid air gave Joe an odd combination of hunger and nausea.  So many tightly packed bodies combined with the smell of the evening meal and the din of so many diners was overwhelming.

Grabbing a compartmentalized steel tray, Joe rode the current of men through the chow line while sweaty sailors, swathed in grease-stained

## My Shadow

aprons, smoked and ladled food as each man passed. A sign on the bulkhead behind the men read: *"Take all you want, but eat all you take."*

Not having much time to negotiate the room, Joe headed toward the first open spot he saw and began his meal by reciting a silent grace in route to a table.

Whether a result of the atmosphere or being onboard a ship at sea, about half way through his meal, Joe knew he was done. Remembering the sign he'd seen, however, and noticing a large sailor monitoring the flow of diners and shouting; "All right! Let's keep it moving. You can look at it later, just get it down and move out!"

Joe quickly shoveled the rest of the food into his mouth, deposited his tray through a slot in the bulkhead and dashed for the door.

Lieutenant Robert Browning was sitting on deck by the hatchway scribbling in a small dog-eared, leather bound book when Joe stumbled out into the cold night for some much-needed fresh air. The meal had not agreed with the young Lieutenant who, until tonight, had never been aboard a vessel larger than an inner tube floating in a pond.

Unaccustomed to the rolling motion of the ship, the vessel was running the anti-submarine maneuvers that would become a frequent occurrence during the voyage. Standing at the port rail Joe looked out over the moonlit North Atlantic, the transport zigzagging its way through the calm sea at twenty-five knots along with the rest of the convoy.

Before the U.S. had entered the war, German U-boats, in an attempt to cut off supplies and starve Great Britain into submission, had patrolled the waters of the North Atlantic with horrific effect. In the early months of 1942, over 87 ships had been sunk off the American coast alone. Fortunately for Joe, during the last year, the U.S. Navy had developed the destroyer escort system that now surrounded him. The dark silhouettes of these sub-chasers had greatly reduced the effectiveness of the German *Wolf Packs* and eased Joe's mind, but not his stomach. With each hard turn of the ship, Joe wondered if being torpedoed would be preferable to the reverberating rumble he was experiencing inside.

Chow had been awful as well as hurried, and this, combined with the motion of the ship had Joe paying a high gastronomic price. He wasn't sure if he should vomit or light another cigarette when he heard someone approaching from behind him.

"Need a light?" came a voice from a young Lieutenant who stepped out from the shadows. "Dammed evasive sub maneuvers are enough to make a guy loose his chow. Not that it would be any great loss. The name's Browning, Rob Browning." The Lieutenant said and tossed his lighter to Joe.

## Robert Brun

He was a thin, young man, barely 20 years old and his mousey brown hair bristled in the light breeze. The pale light of the moon concealed the nausea that he shared, but struggled to conceal.

"You got that right!" Joe replied catching the Zippo lighter with one hand. "I didn't recognize anything they served tonight."

"Nice catch," Rob said dragging heavily on his cigarette.

Joe opened the lighter and, with the unsteady hand of a farm boy at sea, lit his own cigarette. Flicking the lighter closed, he noticed the initials M.C.C. engraved on the cover.

"Belongs to my Mom," Rob said anticipating the question and taking back the lighter, put it in the front pocket of his A-2 jacket, and pulled the collar up against the cold night air. "She gave it to me when I left for flight school. It's her favorite, a gift from her first husband, and she told me I'd better bring it back in one piece or Hitler's wouldn't be the only wrath I'd have to endure. Funny, the way some people say `I love you' these days."

### Abington, England Friday, April 6, 1945

What remained of the daylight continued to fade while Joe stood straining his eyes for any sight of the plane. The war in Europe was now into its sixth year, with the U.S. involved for the last four. As the sky continued to darken in East Anglia, Joe realized he had been fighting it for over a year and a half himself. Had it really been that long? So much had happened and yet, in some ways it seemed like only yesterday. There was the sound of footsteps and Joe turned to see his crew chief, Zeke, walk up beside him and clear his throat.

"Um... don't you think you'd best get some sleep Cap." he said gently, detecting the Captain's concern. "It's getting mighty late and I know you're scheduled to lead tomorrow's mission." He paused. "Don't worry Cap, I'll let you know on the double if anything changes." Joe just continued to stare off into the distance as if he hadn't heard. Zeke was a good man and had been Joe's crew chief ever since his arrival at Abington. Their relationship was a strong one, but it hadn't always been that way.

Technical Sergeant Zeke Schmigrodski was a hardheaded Pole from Detroit and wasn't the kind to make friends easily, but he was the kind to keep them once he did. He was also a man you wanted on your side. An auto mechanic before the war, Zeke could fix damn near anything, and to Joe's constant amazement, regularly did.

Zeke had also become the closest thing to a father Joe had since his dad had died ten years earlier. "Father?" Joe smiled at the thought, Zeke was only 34, but here that made him an *Old Man*. Joe had turned 22 only the day before. He felt much older.

"God, we grow up fast here," he thought.

## My Shadow

"I'm okay Zeke." Joe said crushing out his cigarette on the fender of the jeep and replacing it with a stick of gum, "Think I'll hang around a bit longer."

"Suit yourself, Cap, but you better be alert enough to take care of my plane on tomorrow's mission." Zeke snapped, all trace of empathy now gone from his words and trying to lighten the mood. Giving a nod, the Sergeant turned and headed back across the airfield.

It can seem funny, the relationships you develop serving in the military during wartime. Zeke was as much a friend to Joe as Rob and like Rob, was equally responsible for each of Joe's successful missions. After all it was Zeke and his crew that kept the planes running and in top condition. It was the crew chief's name, not the pilot's that appeared on the plane's manifest sheet. Zeke's was the last face Joe saw before each mission and the first out to the plane after touchdown, but because of the difference in rank, they had to limit their time together to the flight line. Officers and enlisted men simply did not socialize. It was a pity really, but some things you just didn't question, particularly during times like these. Rob's status, however, was different; he was a pilot, a peer... an equal... Joe stopped himself at that thought.

### Monday, November 29, 1943

The North Atlantic crossing continued through a restless evening at sea. The other eleven Lieutenants in Joe's cabin were nice enough guys, but three of them snored and Joe hadn't gotten much sleep.

The following morning, Joe rose early, showered, dressed and climbed up on deck trying to clear his head after a fitful night. Despite the Spartan accommodations, his section, he'd discovered, did have one redeeming factor, fresh-water showers.

This former ocean liner had been converted to the troop transport shortly after the U.S. entered the war and with its reengineering, all of the pre-war niceties had been removed. This included all showers being changed over from fresh to salt water. But apparently, in the haste to do so, Joe's section had been overlooked. As a result, each of the GIs assigned to Joe's deck had been sworn to secrecy by the section officer not to divulge this fact to anyone under penalty of *death*. A minor pleasure, Joe had thought at the time, but one, he would later learn, that could be true delight. The dilemma Joe was experiencing this particular morning stemmed from the salt-water soap each man had been issued for the voyage. Designed to lather in salt water, he had made the mistake of trying to wash his hair with it and was now hoping the sun would dry out the gelatinous mess that had resulted from the attempt.

Looking up from where he sat, Joe noticed Rob at the forward rail staring out beyond the bow of the ship. Seeing a familiar face and feeling a bit lonely on the sundeck, he considered joining his new acquaintance.

Rising unsteadily, having not yet acquired his sea legs, Joe brushed the powdery remnants of the now-dry soap from his scalp and headed toward the bow. All around the overcrowded deck GIs swarmed, talking, sleeping or doing calisthenics. As he approached, Joe could just make out the expression on Browning's face as he sat working in his book. Although not exactly angry, his expression seemed to say *'stay away,'* and with some consideration, Joe decided to follow his instincts. Instead, he went below for another cup of coffee.

For the next eight days and nights, Joe saw little of Lt. Browning and when he did, always with the book. He occupied his time as best he could aboard ship as they made their way through the U-boat-patrolled waters. Few of the men he had trained with shipped out with him, so during the day, there was little to do but sit, wait and think about the future. The trouble was that without anything from his past for comparison, Joe found it hard to imagine what lay ahead.

He found he had little in common with the other Lieutenants he bunked with, all of whom had come from New York City. At times, even verbal communication proved difficult, his southern drawl a constant source of amusement for them.

Reading, gambling, and the occasional fistfight kept some of the troops busy during the crossing, but none of these held much appeal for Joe. Just trying to get fed took up much of his time, and although the meals didn't get any better, at least Joe got better at keeping them down.

There were also numerous ship's pools that anyone could take part in; anyone with money, that is. Who would sight land first? How many days for the crossing? And the disturbing, who and/or how many ships would be torpedoed along the way? Feeling somewhat superstitious, Joe avoided this one in particular.

Nights, however, told a different story. During the long hours of winter darkness, every noise took on a new and ominous meaning.

## My Shadow

Unfamiliar sounds and movements of the ship produced low levels of mild terror that made it hard to sleep. Often, while on deck, Joe would gaze out at the black horizon wondering who might be looking back.

Just after dawn of the ninth day, while finishing the dry toast and coffee that had become his standard morning meal, Joe, heard, then saw the low droning and silhouette of a rotund multi-engine aircraft against the brightening overcast sky. Anxiously looking up beyond the oval smoke stacks as the plane approached, he was relieved to see a Sunderland Flying Boat with British markings pass low overhead rocking its wings. Joe's anxiety lessened and he felt reassured to know that the ship was now within range of the British RAF Coastal Command. They had safely evaded the German Wolf Packs without loss.

A few hours later, pushing his way to the front of the ship, Joe watched the Irish coastline materialize out of the mist like a ghost. It was good to see land again after so many days of nothing but water, sky and clouds.

Rounding the tip of Ireland, the ship steamed into the North Channel, passing Mull of Kintyre and Bennan Head and turned up the Firth of Clyde, heading toward Glasgow. Once again within the lee of land, both the sea and Joe's stomach calmed down.

Joe was surprised by the green and pleasant surroundings. He had expected the bombed out landscape from the newsreels back home. Beginning in 1940 and until the U.S. involvement, Britain stood as the only obstacle to the total Nazi occupation of Europe. This period had indeed been Britain's "Finest Hour," and now, Joe hoped, it might prove to be America's as well.

With land on both sides and the rails lined with gawking GIs, the ship made her way up the Clyde River and eased into her Scottish berth.

Once in port, activity aboard ship turned chaotic. Every last man scurried to get ashore at once, duffle bags slung over shoulders. It took several more hours, but Joe finally made his way down the gangway, and for the first time in his life, set foot on foreign soil.

The dock area was filled to overflowing with arriving troops, but under the directions of MPs barking orders, most unsuitable for mixed company, Joe made his way to the train station.

Despite feeling completely out of his element, he somehow managed, by following the flow of troops, made it into the depot and onto the correct train, where he was lucky enough to find a compartment with a vacant seat.

When the train did pull out of the station heading south to London, Joe spotted Rob pushing his way through the crowded car. This time it was Joe's turn to offer assistance.

## Robert Brun

"Rob... hey Rob! Over here, you can grab a seat here." Joe shouted over the din of the train car while shoving the soldier next to him closer to the wall of the compartment.

"Thanks." Rob tossed his duffle in the overhead and sat down.

The two pilots rode in silence with Joe dozing in and out of a fitful sleep, trying to catch up on some of the shuteye lost at sea. After the rolling motion of the eight days at sea, the jarring *clack* of the train sang a virtual lullaby to the exhausted flier.

The train lurched heavily and Joe opened one eye. Looking across the bench, he could see Rob again; working in the same book he had noticed him with each time he'd seen him aboard ship.

"Writing the Great American Novel?" Joe yawned stretching.

"Huh? Oh. This, nah, just jotting down thoughts and ideas and the occasional sketch" Rob replied holding up drawings of the bow and superstructure of the ship they'd just embarked from. There were sketches of GIs sacked out on deck, the margins around each drawing filled with notations in an illegible hand.

"I hope to work for *Collier's Weekly* one day as an illustrator after all this is over. Writing and sketching helps keep my thoughts straight and my head screwed on right. Been doing it since grammar school and it's gotten to be a habit I guess. Besides, someday I might like to remember all of this for my grandkids." Rob gestured sarcastically around the crowded train car.

"Where you headed?" Joe inquired.

"A place called Abington, wherever the hell that is."

"Oh yeah? Me too. 597th Fighter Squadron, 436th Fighter Group, Eighth Air Force." said Joe, straightening up in his seat and shoving aside the GI who'd fallen asleep.

"Yeah, that's the one. I hear they're getting some new planes to replace the P-38s and Spitfires they've been flying. Pretty good *kite* the Spits, as the Limeys like to say, but not much on range. If we're going to take the fight to old Adolf, we'll be needing something with longer legs. I've been flying P-39s in training I sure won't miss that bird. That thing'll bite ya if you're not careful."

"Logged most of my time in P-40s myself." Replied Joe. "A good solid plane, but it's no match for the 109s and 190s the Jerrys are flying."

The soldier next to Joe grumbled something under his breath and curled closer.

"It's getting kinda crowded in here," Joe said and jerked a thumb at his sleeping bench mate. "You want to get some air?"

"Yeah, and maybe a cup of coffee."

## My Shadow

Joe grabbed his cap and followed Rob out of the compartment. The sleeping Corporal, without waking, dropped with a *clunk* across the bench.

"Huh?" was all Joe could think to say.

The corridor of the train was packed shoulder to shoulder with GIs of every shape, size, description and accent and they all seemed to be talking at the same time.

Pushing through the train, the two pilots reached the dining car, where they met a middle-aged lady behind an up-ended crate serving as a makeshift counter.

"Where can a guy get a cup of coffee around here?" Rob asked.

The woman, after a brief pause, let out a loud guffaw that sprayed a small amount of spittle on the front of his uniform. This was followed by something about *"tay,"* not coffee, in such a heavy accent that to Rob and Joe she might as well have been speaking Greek. At least the other Brits in the car thought it was funny, for they all broke out in gales of laughter and friendly slaps on the back.

*"You Yanks an your sense of umor...Go-On now!"* said the large Scottish sergeant who jovially pounded Rob on the back so hard it almost knocked the wind out of him. Rob winced and looked over at Joe.

"I guess we're not in Kansas anymore, Toto," he said and they headed out of the dining car empty handed.

Pushing farther along, the pair eventually found a spot on a platform between cars. It was cool, damp and breezy, but the fresh air helped Joe shake the cobwebs from his still, weary head.

"So, you from Kansas?" Joe asked conversationally.

"Huh?" said Rob lighting a cigarette.

"Kansas, you know, you said in there that we weren't in Kansas anymore."

"Oh, THAT! No, that's just a line from a movie I saw a few years back. `The Wizard of OZ,' didn't you see it?"

"No, we don't get many films down where I'm from, in Georgia."

"Baldimore." Rob suddenly blurted out.

"What?" Joe asked thinking he'd misheard.

"Baldimore, Merrilyn. That's where I'm from," Rob repeated.

"...And you?"

"Me what?" Joe asked.

"Where-are-you-from?" Rob asked again slowly.

"Oh, uh... Atlanta... Well, outside really, Sharpsburg," Joe replied, wondering if this guy was making fun of him.

"Hmm," was all Rob said in return taking a long draw on his cigarette and blowing the smoke out through his nose.

## Robert Brun

The two men stood silently on the platform for a while, watching the English countryside roll by. Looking out over the patchwork fields, Joe began to wonder for the first time, what really lay in store for him. He hadn't followed the events of the war in Europe during basic training and in flight school. Flying had been far more interesting and exciting.

Watching one small farm after another pass by, it was hard to imagine that he was really in a *war zone* at all. Except for a few thatched roofs or an odd-shaped barn, he could have been traveling across farmland back home.

Joe was about to share this observation with Rob when a growing drone rose above the clacking of the train. Looking up, the two new pilots stared while a flight of B-17 Flying Fortresses passed low overhead. From the ragged looks of the planes, they were returning from another unescorted raid over Europe.

With no fighters protecting the bombers to and from the target, the *Luftwaffe* had been shredding the formations. Evidence of this shortcoming was clearly visible on the planes above them. From where the two men stood, they could see battle damage the Fortresses had endured. Holes in wings, feathered propellers, and missing pieces of fuselage offered all the proof they needed that the *Luftwaffe* was putting up a tough fight.

"Hopefully," Joe thought, "the new planes we'll be getting, can help turn this around."

Coming in low to a nearby airfield, the last Fortress of the group fired off two red flares indicating wounded men on board.

Rob and Joe watched while the damaged plane, with two props feathered, and streaming smoke off the #3 engine, banked to the left then rolled slowly over on its side and spun into a plowed field, erupting into a cloud of thick, black smoke.

## My Shadow

Joe, after a moment, let out a long whistle and looked over at Rob. "You're right about one thing, we sure ain't in Kansas anymore."

For the next three hours the two men tried to stay together amid the dense and shifting population of the train. At last they found a spot and settled down on some crates in a baggage car, their previous seats having been taken over by new recruits shortly after they'd left. All that is; except for the sleeping Corporal who still hadn't moved. Joe half-wondered if he was seeing his first casualty of war.

Settling in for the rest of the trip, Joe again tried to catch up on sleep while Rob went back to his notebook.

### Waco, Texas Thursday, September 16, 1943

The sky was beautiful that day, CAVU, (Ceiling And Visibility Unlimited) and Joe banked the P-40 Warhawk sharply between the rolling cumulus clouds surrounding the training field outside of Waco, TX where he had been completing his fighter training since earning his wings several weeks before. It was hard to believe that so many of his buddies who'd started out together were no longer with the group. Even during wartime, the number of flight cadets that had washed out was still high. This didn't make Joe feel special so much as anxious about what he might do to screw up and get himself sacked as well. Right now, however, he had to put these concerns aside, this was gunnery training and he needed to stay alert.

Turning his head from side to side, Joe noted the position of the three other planes in his flight while he searched the sky for the SBD tow plane they would soon engage.

The SBD known as the *Dauntless* was a dive-bomber the Navy had been using with great success against the Japanese in the Pacific. This particular "war-weary" plane, pulled from active service, had been painted bright orange for high visibility and towed a drogue sock behind it on a cable. This was the target that Joe's flight of P-40s would be using for gunnery practice today.

The process was simple enough. First, rendezvous with the tow plane at 15,000 feet, then circle around for a head-on pass from above. Get the target drogue in your sights and fire.

Each of the four planes was armed with different color-coded ammunition and hits on the target could be determined by the color of the bullet holes in the sleeve-like sock. Simple and in many ways fun, at least Joe thought so, and a far cry from the skeet shooting they'd started out with while learning how to lead a target.

Unlike many of the city boys he'd met, Joe had learned to shoot at a young age using the old bored-out .22-rifle he'd received from his grandfather. Over time, Joe had become pretty good at shooting crows out of his mom's garden, even on the wing. This wasn't that much different.

While the flight stayed in a tight, four-abreast formation, the brightly colored SBD appeared ahead from a cloudbank.

"There she is," Lt. Davidson, their gunnery instructor called over the plane's radio. "Gunnery flight 3-B ready to commence training. Do you copy, target tug?"

Joe was surprised to hear a woman's voice respond.

"Target tug, roger. You may begin when ready." Everyone was doing his part, even when *he* happened to be a *she*.

"You've all been through this before, so let's see what you've got." Lt. Davidson looked over at the three pilots, gave a nod and dropped his plane back out of the formation.

Joe was number three in the group and followed Flight Cadets Dan Clark and Ben Howard who, now pulled their planes out in front of the SBD to begin their passes at the target. Joe checked the position of the other two planes and toggled the switches that activated the N-3B reflector gun sight. The bright targeting reticule or *pipper* illuminated in the reflective glass and the six 50-caliber wing-mounted machine guns charged for firing.

Flight Cadet Clark, first up, peeled out of the formation to begin his pass. Approaching from twelve o'clock high, puffs of smoke appeared and shell casings began to drop from Clark's wings. The white streaks of the tracers showed that Clark's aim was a bit high as he banked hard to the left after his attack run. Joe was certain he could do better. Next up was Ben Howard.

A funny little guy, Ben was built like a fireplug, standing barely 5'5-1/2" (The Army's minimum height requirement was 5'6," and Ben, rising up on his toes during his physical, had hoped the Army medics wouldn't notice, and they hadn't.)

Short as he was, Ben was determined to make it up in *piss and vinegar* for anything he lacked in stature. Since his posting to Waco, Joe

## My Shadow

had seen Ben in no fewer than four fistfights, each with much bigger guys and in each case coming out the victor. Ben didn't start fights he finished them. Joe both admired and respected Ben. It was an unlikely friendship though: Like *Mutt & Jeff*, Joe, the tall, thin and more reserved to Ben's short, stocky outgoing personality. The two couldn't have been more different, and these differences brought the two together.

Ben Howard also flew like he fought, all out, *balls to the wall*, never afraid to push both man and machine to the limit. Ben was always the most aggressive pilot of any flight. Diving steeper, flying lower and pulling greater g-forces than any other cadet, always getting in closer to the target and making every round count. Joe had learned a lot from just watching Ben's style during their training flights.

Though strictly against regulations, Joe and Ben had often mixed it up in mock dogfights when flying patrols and when they did, Ben always got the drop on him. Joe was glad Ben was on his side. He wasn't sure, however, if he'd ever have the nerve to actually use some of Ben's tactics. The guy could really fly!

Joe watched anxiously as Ben rolled his P-40 over into a diving-attack position. Ben, in true form, had chosen a steeper angle, coming in from above in an almost vertical dive.

"Too steep?" Joe thought watching his buddy start the attack run. With Ben's plane now pointing straight down, Joe could see multiple hits on the drogue. He was really peppering the target with his bullets.

"Way to go Ben. Old Jerry won't have a chance with you on the job." Joe thought to himself, admiring his buddy's shooting skills.

Then, in the next second, everything changed. Apparently misjudging his angle and high speed, Joe watched Ben cut in too close to the target drogue between the tow cable and the plane. At the speed he was traveling, it was impossible to maneuver and clipping the cable with his port wing, it sliced the outside four feet of the Warhawk's wing completely off. While Joe watched in horror, his buddy's P-40 flipped over into a rapid counter-clockwise spin and careened toward the desert below.

"Bailout! Bailout!" Joe screamed into his oxygen mask, but he knew at the speed with which the plane spun, there was no chance of seeing a chute.

With a screech of steel and a violent lurch, Joe was shaken awake, the train braking into the London station.

"Who's Ben?" Rob asked, looking up and closing his notebook.

"Huh? Oh, uh, a buddy of mine from flight school." was all Joe said in reply.

Rob didn't inquire further. He could see the beads of sweat on Joe's forehead.

# Robert Brun

## Chapter 2
## Abington

*Friday, December 3, 1943: Arrived at the base today after a train ride from London. Met up with Lt. Dyer. Seems like a good egg. Saw our new planes. Mustangs! Can't wait to try one on for size - R.B.*

A light rain fell as the train came to a stop at the London depot.

"This is our stop," Rob said struggling to his feet along with what seemed like every other soldier aboard the train. "From here we go by truck... uh, lorry."

Joe grabbed the side of the crate he had been using as a cot and rose unsteadily to his feet. He still carried the image of Ben Howard's plane disappearing into a ball of flame as he hoisted his duffle bag onto one shoulder and followed Rob and the rest of the GIs out into the morning drizzle.

The London scene that now surrounded him was not at all what he had expected, and was a far cry from the pictures in the geography books he'd studied in school. In the morning light Joe could just make out the silhouettes of bombed-out buildings that surrounded the train station.

The *Luftwaffe* had been flying nightly bombing raids over London ever since the RAF had put a stop to Hitler's invasion plans back in September of 1940, bombing mostly out of spite Joe supposed. The odd thing was, despite the ruins and burned out buildings, you would never know that war raged barely eighty miles away across the English Channel.

Outside the station, people bustled about as though this were a regular workday in any other city. Men in suits carrying umbrellas, delivery trucks and busses, and even women, some with children in tow all went about their business as though the rubble from the bombing were just so much street litter.

"No wonder the Jerries lost the Battle of Britain" Joe thought, "These people are tough."

## My Shadow

*"You one of dem new lot from da States, Yank?"* someone with a very thick Cockney accent addressed Joe while he stood there taking in the broken skyline.

Lowering his gaze, Joe saw a tall, ginger haired man in an RAF uniform standing near a canvas-covered truck.

*"Better get yeself up into dis eer lorry wit da rest o your mates or you'll quite rightly be lef be-ind and we're not sheduled to pick up another lot for o-er a fort-night."*

Glancing beyond the gentleman, and not entirely sure what he'd just heard, Joe saw Rob, already in the back of the truck beckoning for him to climb aboard.

"Joe!" Rob hollered, "Never mind what he said just throw your gear back in here with the rest of us *Yanks*."

Tossing his duffle up to Rob, Joe scrambled into the back. The truck pulled out of the station as the first rays of the new day's sun broke through the clouds over the bombed-out buildings of London. Joe felt very far from home.

### Abington, Friday April 6, 1945

The sun had now set as Joe continued to stand there at the edge of the runway. There was still no sign of his wingman's plane, and it had been over three hours since the last of Red flight had landed. Where was Rob? For the first time since they'd been flying together, the two had gotten separated, not that it really surprised him, given what they'd encountered on that day's mission over Germany. Joe had never seen the *Luftwaffe* put up a fighter defense like the one they'd run up against, particularly after the lull of the last few weeks. It was a wonder anyone had made it back at all.

The mission had been a long one, made more so by that morning's weather delay and it had gotten off to a bad start in more ways than one.

Conditions over the target area were reported daily by a weather plane, usually an RAF deHavilland Mosquito or a modified P-38 called an F-5, painted light blue to blend with the sky. These unarmed weather-reconnaissance planes relied solely on speed and altitude to get in, assess the weather over the target, and get out without being intercepted. That morning the report had been ten tenths cloud cover over the target. The result being, the fighters of the 597th hadn't gotten airborne until afternoon when the weather finally started to clear.

Despite the delay, the mission had been a milk run most of the way in with no enemy fighters spotted until well inside the German border.

Ever since D-Day, the appearance of the *Luftwaffe* had been at best, sporadic. In fact, for the last twelve missions, Joe hadn't seen a single enemy fighter. Maybe he'd gotten rusty. As much as he'd like to think this meant the Germans were licked, he knew that this was just the kind of thinking that could get you killed.

For weeks Allied Bomber Command had been planning this mission, a heavy strike deep into Germany. Because of the continual bombing, the Germans had moved the industrial center of the *Reich* out of the Ruhr Valley toward the Czechoslovakian boarder. This made for very long mission days.

It was an incredible sight flying above the endless stream of heavy bombers that spread out before him and to the horizon. Well over eight hundred B-17s and B-24s and an equal number of escort fighters now made up these bombing missions, a far cry from the smaller groups the 597th had escorted just a year before. The constant hum of German radar in his headphones told Joe the Jerries had already picked them up and probably figured out where they were headed, but still there was no sign of the *Luftwaffe*. Four hours into the mission and practically at the I.P. (Initial Point of the bomb run) the German fighters made their move.

They came *en masse*. The *Luftwaffe* called it a *Schwarm*, hundreds of planes in dozens of groups materialized from behind the dark clouds ahead of the bombers. The sight reminded Joe of the morbid image of flies ascending from the carcass of a dead cow he'd once seen on his uncle's farm.

Bf 109s, FW 190s and even Bf 110s as far as the eye could see all armed to the teeth and all headed right for the bombers and Joe, Rob and the rest of the 597th were right in the thick of it. Looking out at that many enemy planes, it was hard to believe eleven hours earlier, this day had started just like every other mission day.

At 05:30 that morning, Joe awoke and crawled out of his cot unwrapping the five scratchy wool G.I. (Government-Issue) blankets he'd rolled into his own personal cocoon. Wrapping up this way allowed him to trap his own body heat and keep warm enough to sleep. The canvas cots provided by the Army offered little in the way of insulation, and the trick, he'd learned, was to put most of the blankets beneath you.

It was still dark outside, and the fire in the small wood stove used to heat the Nissen hut had long since gone out. As pilot officers, Joe shared his particular hut with only three other men, unlike the ground crews and enlisted men who slept in barracks'. Capt. Altman, one of Joe's hut mates, was making a hell of a racket stuffing kindling and pages from *Stars and Stripes* through the small open grate trying to get a fire burning again. Although Joe was appreciative of the Captain's effort, all Altman seemed to be accomplishing was filling much of the room with a thick cloud of white smoke and singeing the hair from his knuckles while swearing a blue streak.

## My Shadow

Joe ignored the coughing and grumbling from the other bunks while he pulled on trousers and boots over his long johns, finding the strength to stand upright.

Even though he had just celebrated his 22nd birthday he felt more like an octogenarian and the excessive drinking he'd done the night before certainly hadn't helped. Every movement Joe made was accompanied by the now familiar *pops* and *cracks* of his joints that resulted from too many of these cold, damp English mornings. How the Brits had endured this climate and thrived for centuries, he would never understand.

Returning from the latrine at the end of the hut where he had washed up and shaved (so his oxygen mask wouldn't chafe too badly) he ran into the CQ (Charge of Quarters) with his flashlight. It was this Corporal's thankless job to wake the men scheduled for that day's mission.

After checking off his name on the clipboard indicating he'd been awakened, Joe pulled on his wool topcoat and cap and made his way to the door of the hut. On opening it, he cringed, instantly assaulted by a blast of cold, damp air from the pre-dawn morning, and with a shiver, Joe exited the relative warmth of the hut. Balancing along the wooden planks that covered the mud, he headed toward the mess hall for breakfast.

In the predawn light, he no longer took much notice of the steady drone of radial engines above the cloud cover in the overcast. The slower bombers always took off well before the escort fighters.

"Poor bastards" Joe thought, "and we think we have to get up early." But Joe knew that his plane, and the rest of the Little Friends would be catching up with the *heavies*, soon enough.

Bicycles were the standard pilot transportation around the base and for short trips into town and the pub. Unfortunately, the 597th was one of the few remaining 8th Air Force's fighter groups still flying from the older RAF grass field they'd adopted over a year ago and the mud that had formed on the perimeter road from that night's rain was so thick that it would only clog a bicycle's wheels.

Joe slogged his way through the ankle deep muck to the Jeep where Captains Altman and Taylor were already waiting huddled together in the back for warmth, their caps pulled down over their ears and collars turned up covering their pale faces. Their thousand yard stares evident even in the pre-dawn gloom.

Joe hopped into the Jeep's front seat and waving away Altman's offer of a cigarette, bounced across the muddy field to the mess hall.

The three pilots rolled out of the Jeep and over to the mess hall, past the maintenance bays where Joe could hear the metallic clanking of tools, and shouts over the run-ups of their Mustang's engines. The mechanics were always busy preparing the fighters for the day's missions and Joe felt a mild twinge of guilt mixed with pride as he walked by.

## Robert Brun

"Don't these guys ever sleep?" He'd guessed not, or at least not that he'd ever seen. "Maybe during the missions?" What Joe did know was that Zeke and his crews had kept both his planes running flawlessly since his arrival and he'd never had to abort a mission.

After so many missions a pilot gets into a routine, and like many of the veterans, Joe no longer bothered with a real breakfast on days he flew. Even the fresh eggs provided on mission days no longer held much appeal for him, mostly because he knew he was likely to lose them on his way to the flight line. *Pre-flight stomach* is what Joe called it, all too common among the pilots and as a result; he always flew more comfortably on a light breakfast. He'd eat after he returned.

Entering the mess hall and bypassing the vats of congealed porridge, greasy bacon and the ubiquitous SPAM, Joe filled the cold enameled mug from one of the multiple pots of coffee.

"It's certainly not like mom used to make at home," he thought as he had every morning since his arrival while looking down at the steaming brown liquid and grabbing a piece of dry toast from a tray, "but at least it's hot." The warmth felt good on his cold, stiff fingers.

Making his way across the mess hall, Joe spotted Taylor, already seated and stuffing his face with food. The sight made Joe mildly nauseous, there was no sign of his wingman. Joe wasn't surprised.

"Joe!" Taylor said cheerfully looking up. "I see you slept in again today. How's our Squadron Leader this fine spring morning?"

"Can it Taylor, I'm not even awake yet. Has anybody seen the sugar?"

Like a bartender serving up a beer in a Western movie, the sugar container appeared, sliding down the table and coming to a precise stop right next to Joe's mug.

"Thanks." and without looking up, Joe poured a long stream of white granules into his coffee.

"Where do you think we're headed today? Wow! Did you hear those bombers?" 2nd Lt. Matt Jefferson asked with the excitement only a raw recruit could bring to a day like this.

Jefferson, one of the New Guys, hadn't yet learned the Cardinal Rule, the one that said *"Thou shalt not ask about a mission prior to the briefing."* Jeez, fighter pilots were a superstitious lot, but then again, after more than eighty opportunities to get oneself killed I guess anybody would be, the odds had not been in Joe's favor for sometime now. Everyone at the table ignored the question and Jefferson put a lid on his enthusiasm.

Joe had never liked this kind of behavior, especially toward someone as inexperienced as this new second Lieutenant, but it was a tightrope-tough balance a pilot had to maintain about who to befriend and who to keep at a distance.

## My Shadow

The guys you came in with, that was easy, they were your buddies, whether you liked them or not. They knew what you knew and had been through what you'd been through it formed an unbreakable, impenetrable bond. When one of these guys *bought the farm*, so called because of their $10,000 USAAF insurance policy courtesy of Uncle Sam, it could hit you real hard, perhaps strangely so. Not so much the first time, that one was such a new shock your mind just blasted it away and retreated into itself, but the second death, that was the worst and it didn't even have to be someone close. Then, not only did it hit you, but the first one also came back in spades and the two memories combined to beat your emotions senseless.

Joe thought of that first fatality. It had been his first mission and he hardly even knew the guy. One second he was there, the next... it had been that fast. In the middle of a mission, however, you don't have time to grieve or even think, you just do, and try to stay alive. After you land, *that's* when the reality of the situation hits home.

"This is war and people are going to get killed," Joe remembered thinking as he'd shut down the Mustang that first day "and the next time, it might be me!" He sipped at his coffee and thought about Rob.

From that point on, after that first death, the mind comes up with strange ways to cope. Anger, fear, elation, relief, guilt and a strong need to distance oneself from others, all played their part. All that emotion is rather hard to deal with until you find a strange, cold-hearted balance you can live with.

The odd thing was, as bad as it could sometimes be, it was only troublesome while on the ground, between missions, when there was too much time to think. Some of the guys took it harder than others and often resorted to a bottle to drown their emotions, but for Joe, once back in the cockpit and airborne the confused, ambivalent feelings all disappeared. No fear, no guilt, not even anger, just a job to do, and over the last year, Joe had learned that job very well. Maybe it was because he didn't have time to think about anything else, but it had gotten so Joe started to feel out of balance unless he was strapped into his Mustang, breathing through an oxygen mask and flying five miles above the ground. So, you separated yourself from the other men, especially the new guys because many of them never even made it through those all-important first five missions.

Funny thing was though, despite all of this, somehow Rob had always been different. Sure, they had arrived together and many thought of him as an *odd duck*, but for all his quirks, Rob wasn't pushy or loud like some of the other pilots and never seemed to be trying to prove anything. He kept his distance, and yet he was there when you needed him even if maybe you didn't appreciate it at the time. Joe winced at that thought.

## Robert Brun

Rob had always been different and at first it was hard to tell if he even liked you. Not that he was mean or anything like that, it was just that he didn't seem interested, or maybe it was because he always had his nose buried in that book of his. It's hard to disturb someone who's working.

Initially, Joe hadn't thought much of Rob's ability as a pilot. At first, he'd even thought maybe the guy was a screw-up and would be sent back stateside for refresher training. It hadn't occurred to him at the time that Rob was going through exactly the same first mission jitters Joe was wrestling with, and to make matters worse, in a new fighter they'd been flying for just two weeks.

### Friday, December 3, 1943

Arriving at the base that first day was much as Joe had expected. The three-hour lorry ride from London had been far less than comfortable. Not only was it unnerving to be driven on the left, the *wrong side* of those narrow country roads, but squeezed in among all those other GIs, none of whom had showered since leaving the ship, things were getting a bit ripe. So it came as a great relief when they turned into the base at Abington in East Anglia and the driver lowered the tailgate.

Groaning and stretching, the men piled out and looked around. The sun was forcing its way through broken clouds while fighter planes came and went overhead. Joe grabbed his duffle and followed the rest of the group into a small building where they were instructed to take a seat.

Unlike the train down from Glasgow, this setting was exactly like Joe had seen in the movies. A small, smoke-filled Nissen hut jammed with chairs and a large covered board at one end of the room containing the information for that day's mission. Joe's heart rate jumped a little. He had finally arrived.

Glancing around the room, a few of the faces were familiar from the train and there were others he didn't recognize. Way over in one corner, Joe spotted Rob, working away in his book, but this time in a manner that seemed to indicate he was trying to minimize a nervousness that he too was feeling. Joe lit a cigarette and waited.

Some short time later, an unidentified captain strode in and addressed the assembly. Joe and the others received their hut assignments and other necessary information about the base and filed out blinking into a now bright sunny afternoon.

As the men made their way around the unfamiliar airbase, Joe glanced over to see Rob standing in front of one of the open maintenance bays. Curious to see what had caught Rob's attention, he walked over and stopped dead in his tracks. Parked inside the hangar was the most striking fighter plane Joe had ever laid eyes on. A brand spankin' new P-51B

## My Shadow

Mustang, the first one he'd ever seen, getting a coat of olive drab paint. Joe dropped his duffle and let out a long whistle.

"You can say THAT again," was Rob's reply. "I think I'm in love."

"Hey Zeke!" One of the maintenance crew hollered to a sergeant standing beside the plane. "Get a load of the two virgins!"

The sergeant looked up for only a moment, smiled, and shaking his head, spat on the ground and went back to his work.

"Just get that bird painted by 04:30 tomorrow or I'll have your hide." was all he said.

"Have you ever flown a Mustang?" Rob whispered to Joe.

"No" Joe replied.

"Me neither. I just hope they're as easy to fly as I've heard."

The next morning, Joe got his chance to find out.

# Robert Brun

## Chapter 3
## Training

*Saturday, December 4, 1943: Taking up the new plane for the first time today at 14:00. Time to see if my training is worth what Uncle Sam paid for it - R.B.*

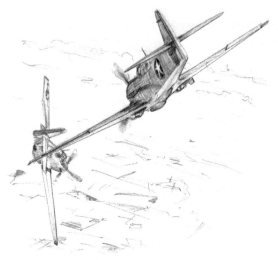

Early the next day, Joe and the rest of the new pilots received their plane assignments and were given basic instructions on the operation of the P-51B Mustang. Joe had spent most of the previous day and late into the evening studying the flight manual he'd been issued. By morning, he hadn't gotten much sleep, but his excitement more than made up for his fatigue.

Fortunately for Joe, the North American P-51 had been designed by the same company that had built the Advanced Trainer-6 or AT-6 *Texan* he'd flown in flight school. He was also relieved to discover most of the controls were just where he expected them to be, including the *relief tube* under the seat.

The maintenance crew's Corporal helped Joe strap on his parachute and shoulder harnesses, while Captain Petersen gave Joe some last minute instructions.

"She's lighter on the controls than the P-40 and doesn't pull quite as much on takeoff, so watch your rudder trim. Once the wheels and flaps are up, get some altitude and put her through a few turns until you get the feel of things, then circle the field at 8,000 feet and wait for the call from Major Higgins."

## My Shadow

"Major Higgins?" Joe asked, confused, but before he could get an answer, the Captain slammed closed and secured the canopy with a slap that signaled Joe to start the engine.

Initiating the Mustang's starting procedure, Joe ran through his mental checklist:

Flaps: *15%*
Carburetor: *set to normal*
Coolant: *auto*
Rudder Trim: *five degrees right*
Elevator: *four degrees down*
Ailerons: *neutral*
Fuel Mixture: *idle cutoff*
Throttle: *one inch*
Fuel booster: *on*
Left and right magnetos: *set to both*
Fuel flow: *select left wing tank*
Brakes: *set*
Batteries: *on*
Generator: *on*

With a wave, the ground crew's Corporal signaled *clear* and Joe pressed and held the starter button. The big four bladed Hamilton Standard propeller slowly began to turn and after five revolutions, the powerful Merlin engine caught with a bang. White smoke belched out of the twelve exhaust ports and Joe switched the throttle mixture to the *run* position.

Waiting anxiously while the oil temperature rose to the limit for proper warm up, the engine of the Mustang coughed and popped, bucking like its equine namesake.

Signaling the Corporal to remove the wheel chocks and releasing the brakes, Joe taxied the unfamiliar plane out onto the end of the airfield using the rudder pedals to weave the plane back and forth in order to see around the long engine cowling that blocked his forward view. After running the engine manifold pressure up to 30 inches, Joe closed the radiator vent flap, and locking the tail wheel, adjusted the propeller pitch preparing the fighter for takeoff.

Advancing the throttle to full power, the Merlin engine smoothed out and the Mustang, unencumbered by the weight of its usual armament load, leaped forward and raced down the field. Reaching the 100 mph takeoff speed, the tail came up and with a slight pull back on the control stick; man and machine were soon airborne.

Once free of the ground, the plane banked sharply to the right as Joe over-corrected, compensating for the propeller torque. The Corporal had been right; she didn't pull as much on takeoff, far less than the P-40. Seconds later, Joe operated the lever that retracted the landing gear and

## Robert Brun

banking off to the left, executed a climbing turn while continuing to familiarize himself with the flight characteristics of the P-51B.

"Holy Cow!" Joe thought, amazed at the power and responsiveness of the Mustang compared to the P-40s he'd been flying. "With a plane like this, they'll be no stopping us."

Elated, Joe soared all over the English sky, executing rolls, loops, split-S, and Immelmans, making mental note of the plane's acrobatic and stall characteristics. There was no doubt, the Mustang was a handful, but like any thoroughbred, it would do whatever you asked of it, if you were willing to put in the effort, and this plane was well worth that effort.

Still admiring the plane's performance, Joe recalled the Captain's instructions so climbing to 8,000 feet, he leveled off easing the throttle back to cruise and fine-tuned the fuel mixture and prop pitch while slowly circling the field waiting... for what, he wondered?

After about 10 minutes of impatient circling, Joe noticed a second plane off to his right. Expecting another Mustang, he was confused and then surprised to see the familiar silhouette of a P-40 climbing about 300 yards ahead of him. Assuming it to be a training flight or one of the maintenance crew out for a joyride in one of the bases retired Warhawks, Joe relaxed, then was stunned when the P-40 turned directly into his path and flashed by, missing him by only a few yards.

"What the hell?" Joe shouted whipping his head around.

Pulling hard on the control stick with both hands so the Mustang would follow his gaze, the P-51 made the 180-degree turn with some effort, almost stalling when the port wing dropped unexpectedly.

Shaken by the close call, Joe recovered his composure only to see empty sky. The mysterious P-40 was nowhere in sight. Trying to figure out what was going on, Joe was again startled when the same P-40 raced by overhead, so close he was buffeted by its prop-wash when it peeled off to his right.

For the next ten minutes, Joe wrestled the Mustang around the sky almost losing control of the fighter several times. Now breathing hard, he could feel beads of perspiration forming beneath his flight helmet while fighting to maintain control of himself and the new fighter. All the while, the strange P-40 continued to dog his every move, anticipating his maneuvers and heading him off at every turn. Joe was reminded of the mock dogfights he and Ben had played at during training, but this was far more dangerous. What was going on?

Coming out of the sun and several times, appearing directly behind him, the Warhawk seemed to be everywhere and yet nowhere at the same time. No matter what Joe tried, the mysterious plane was always one step ahead of him. Why didn't it show itself, always sneaking up like it

# My Shadow

did? How could he be expected to get the upper hand if the other guy didn't fight fair?

Frustrated by the overwhelming series of mock-attacks Joe, now completely exhausted, tried to break off from the contest without luck. Just who was this *yahoo* pilot anyway?

Irritation gave way to anger that one of our own guys would have engaged him in such a foolish and dangerous manner, but this did little to ease Joe's annoyance at being bested his first time out. Seconds later, a voice broke in over the radio.

"Lieutenant Dyer. Proceed to heading 233 degrees and return to base, on the double."

Still unclear what was going on, Joe acknowledged the order with a curt reply and returned to the base landing the plane as instructed and taxied the Mustang back to the hardstand. Looking across the perimeter track, Joe spotted the P-40 that had been dogging him parked nearby. Still fuming, he shut down the Merlin, pushed the canopy open, and climbed out of the cockpit. Jumping down off the wing Joe stormed over to the other plane anxious to confront the pilot who had just endangered his life, not to mention a brand new and very expensive piece of Army Air Corps flying equipment.

"Hang on a minute, Lieutenant," said the line chief, intercepting Joe as he approached the plane. "I like to introduce you to *Major* Higgins."

Joe stopped in mid-stride as the P-40's canopy slid back and Major Higgins, looking like something out of *Terry and the Pirates*, climbed out onto the wing. Major Higgins was a short, stocky man, an unlit cigar clenched tightly in his teeth and a stern look on his unshaven face.

Stunned by the sight of this unfamiliar officer, all Joe could manage was a sloppy salute, and turning his face away in embarrassment, noticed the white-stenciled WW painted on the tail of the Warhawk.

"War Weary" Joe thought with a sigh. "Not even good enough for combat duty and it's just flown rings around me!"

The Major climbed down from the wing and came over to where Joe stood, the tip of his cigar so close Joe could smell the damp tobacco.

"Lieutenant," the major said, getting Joe's full attention and looking him straight in the eye despite his shorter stature.

"You are flying the most advanced fighter aircraft the U.S. Army has at its disposal. A plane that, we hope will turn the tide of the air war over the continent. May I remind you Lieutenant, it's not the plane you're flying that makes the difference, it's the quality of the pilot. As the Baron Manfred Von Richthofen would have put it, 'It's not the crate; it's the man sitting in it.' Remember that, Lieutenant, and you'll live a lot longer!"

## Robert Brun

With that being said, Major Higgins turned with a grunt and walked away, leaving Joe standing alone and feeling about two inches tall.

Still embarrassed by the morning's event, Joe spent the rest of the day reviewing all available information on the Mustang. It had been hard to admit it, but he now knew, he had a lot to learn.

Later that evening Joe cautiously mentioned the incident to Rob at dinner.

"Have you met Major Higgins yet?" Joe inquired sheepishly.

"HAVE I!" was Rob's quick reply, slamming down his fork with a bang, "That guy really cleaned my clock this afternoon. Boy, oh boy! I thought I was a real *H.P.* flying my Mustang until the Major flew rings around me in that beat-up old P-40. Sure brought me down a peg or two."

"I think that was the idea." Joe said.

All the replacement pilots were unusually quiet that evening.

For the next two weeks, Joe and the rest of the new pilots met daily for pilot briefings with Major Higgins. Joe flew the P-51 as often as possible, listening with a new found attention to what the veteran pilots, Zeke and ground crews had to say about the flight characteristics and tactics best suited to the Mustang for combating the *Luftwaffe* fighters. Joe even studied WW I ace Oswald Bölcke's *Dicta* about aerial combat:

1. Try to secure the advantage before attacking.
2. Whenever possible, keep the sun behind you.
3. Always carry through an attack when you have started it.
4. Fire only at close range and only when your opponent is properly in your sights.
5. Always keep your eyes on your opponent.
6. Don't let yourself be deceived by a ruse.
7. In any attack it is essential to assail your opponent from behind.
8. If your opponent dives on you, do not try to evade, but turn to meet him.
9. When over enemy lines, never forget your own line of retreat.
10. When the fight breaks into single combat, take care that several planes do not go for one opponent.

Joe had been relieved to learn that he and Rob were not the only pilot subjected to one of Major Higgins' little *flights of fancy* as they had come to be known around the base. As it turned out, this was standard procedure for all new pilots fresh out of flight school and a bit too *gung-ho* for their own good. The base commander, Colonel Tomlinson had initiated this orientation procedure about three months prior to Joe's

## My Shadow

arrival, and though a humbling experience, it had considerably cut down on mission casualties since then.

At 06:30 the following Monday, Joe along with the rest of the squadron were again assembled in the briefing room, smoking and restlessly waiting for the base commander to arrive.

Colonel G.W. Tomlinson had been the base commander of the 597th Fighter Group for almost a year when Joe arrived. Promoted up the ranks from 2nd Lt, he had been a member of the original *Eagle Squadron* that flew with the British during the Battle of Britain. Some of the other pilots had also flown with this man back when the USAAF was still borrowing Spitfires from the British in early 1942.

During the transition to the P-38 Lightning, Tomlinson had been shot down twice over occupied France, each time managing to make it back in one piece, although the second time it had taken over two months. Rumor was that he'd even worked with the French Resistance and had stolen a Bf-109 for his escape.

Unable to get the German Fighter's radio working, he was nearly shot down over the Cliffs of Dover by a patrolling Spitfire pilot. Not sure how to identify himself as an American and unable to figure out how to lower the landing gear, the universal sign of surrender (unknown to Tomlinson, the 109 used different control levers to raise and lower the gear) he rolled the plane over on its back and flew the last ten miles upside down. The befuddled RAF pilot, curious to learn who this inverted nut was, escorted him to the nearest aerodrome where he bellied the plane in. Now, because of his detailed knowledge of the French Underground, regulations prohibited him from flying combat, less he be shot down and captured again, something that had not made him happy.

Highly respected, and some would say even liked by his men, Tomlinson was one of those commanding officers who instinctively knew when to cite regulations and when to toss the rulebook out the window. He knew what it was like to fly and fight in actual combat, not just in theory. Because of this, Tomlinson had earned the men's unwavering loyalty. They trusted this man and his leadership.

With another gust of frigid air and a fair amount of swearing, the door to the briefing room opened and was immediately followed by a brusque *"A-ten-Hut!"*

Chairs slid across the floor as the men stood at attention. The Colonel strode in and crossed the room to the curtain-covered wall where he turned and faced the assembly. Scanning the room of pilots as if sizing them up, he cleared his throat and addressed the men in a deep baritone voice that immediately got Joe's attention.

# Robert Brun

"At ease, men." and the room instantly returned to a more casual atmosphere.

### Thursday, April 6, 1945

Joe smiled to himself, standing there on the darkening field and thinking back to that first mission briefing. He had been one of the *new guys,* wet behind the ears and about as green as a hickory switch. If not for Rob's blunder that first day, he probably would have been a lot more nervous than he had been.

It was said by the veteran pilots that all you had to do was fly those *first five* missions and you'd be okay. Of course what that really meant was that if you managed, five times in a row, not get yourself killed, then you'd have naturally learned enough to keep yourself alive. Like everyone else, as Joe later came to learn, the first five were the hardest and for this first mission, he worried most about his reaction to actual combat.

Joe had been through the maneuvers and the training with Major Higgins. He'd studied all the diagrams and even endlessly questioned the pilots who'd come back from their missions, but that was all just theory and stories, today was going to be for real. Real combat, real bullets and a real enemy trying to *really kill you!* And each one of the *Luftwaffe* pilots, he knew with grim realization, was far more experienced then he was. Would he be able to do his job or would he panic or worse, would he just freeze like he'd heard some pilots did? This was something no one could train for, the moment of truth, the *Baptism by fire* and it bothered Joe a great deal because, ready or not, in about one hour, he'd have his answer.

That first mission had been easy based on what he now knew, but at the time it felt like he was preparing to take on the entire German Air Force all by himself.

"You'll be flying air cover for a flight of British A-20s, *Havocs* attacking a base in Amiens," the Colonel continued, his aide drawing back the curtain uncovering a wall-sized map. Red yarn showed the routes to and from the mission. With a pointer, Tomlinson indicated where the fighters would form up, where to rendezvous with the bombers and just what the objective of that day's mission was. Looking around, Joe noticed some of the other pilots jotting down notes on their hands. He reached for his own pen.

The *Luftwaffe* had been using an airfield outside Amiens as a staging and repair station for the Focke-Wulf 190s used by the Yellow Nosed JG-26 or *Abbeville Boys,* the *Luftwaffe's* elite fighter group. Destroying this repair facility would cut off the supply of replacement planes to these pilots that were raising hell with Allied bomber groups heading to and from the Continent. Reconnaissance photos had shown

## My Shadow

that, because this was a secondary base, the anti-aircraft defenses would not be heavy.

Joe nerves were such that when the order came to synchronize watches, he almost pulled the stem completely out of his wristwatch.

"That's your mission, men, keep close to the bombers and give the Jerries *what for*. We've got a few replacements coming along for the first time on this mission. To you I say, stick with your flight leaders, keep your eyes and ears open and follow what you've been taught. You're with a good group. You'll learn and that'll help you stay alive." With that the Colonel turned to leave.

"*A-ten-hut!*" barked the Colonel's aide, and as Joe and the rest of the men stood to attention, a loud clatter came from somewhere in the back of the room.

Looking over his shoulder, Joe could see Rob lying flat-out on the floor thrashing about and desperately trying to regain his footing. The Colonel stopped briefly as he approached the door of the hut and glanced over at Lt. Browning lying amidst the toppled chairs. Rob, apparently unable to think of anything better to do, gave the Colonel a sharp salute while balancing up on one elbow. Tomlinson fixed his gaze briefly on the prone Lieutenant, then exited the room, his aide followed, shaking his head. Second Lieutenant Robert Browning let out a loud groan and fell back to the floor with a thud.

# Robert Brun

## Chapter 4
## Baptism

*Sunday, January 16, 1944: First mission! First crack at the Jerrys. No more rehearsals, this is the real thing! - R.B.*

Later that morning while Joe sat in the cockpit of his Mustang, scanning the instruments and audibly checking the sound of the idling Merlin (had it always been this loud?) it seemed as if everything he had ever been taught about flying and combat was running through his head at the same time. All the information was there, that he was sure of, but would he be able to sort through it and get it in some usable order when it really mattered?

Lost in thought, Joe was snapped out of his daze by the growing roar of another Merlin engine coming from behind his plane. Looking up into the rearview mirror he had caught a glimpse of a red and white spinner closing fast from behind. With a purely instinctive reaction, Joe released his brake and throttled his Mustang forward enough to avoid the spinning propeller blades behind him saving his rudder.

"Damn! I'm not even airborne and already I've got a bandit at six o'clock!" he thought. Joe craned his neck to identify the joker who'd almost clipped his tail feathers, even before his first mission. Looking back, a second Mustang with the code M7-B turned right and came to a stop in its proper taxi position next to Blue Flight's leader.

"Browning again," Joe thought with a shake of his head. "That guy's gonna get somebody killed and probably himself as well!"

Joe and Rob had been assigned to Capt. Maxwell Adams and Lt. Charles Perkins making up the four planes of Blue Flight for this first mission. Capt. Adams was a 597$^{th}$ veteran of over 40 missions with 2 confirmed kills, 1 damaged and 2 probables. He was Blue Flight's Leader

## My Shadow

and Joe was flying as his wingman. Glancing over, Capt. Adams gave a quick thumbs-up to the rookie pilot as if to say "Just stick with me, do your job and you'll be fine." Seeing this made Joe feel a bit more relaxed.

Running up the Mustangs' powerful twelve-cylinder engine filled the cockpit with the familiar roar and the planes rolled forward while the ground crewmen, sitting on the wingtips, signaled Joe and the other pilots directing the fighters down the taxi way and into position for takeoff. Joe advanced the throttle and pulled out next to Captain Adams' plane.

Joe and the other Mustangs were to take off and form up with the rest of the group at 5,000 feet above the field before heading out across the Channel and locate the bombers at the rendezvous point. Because of the planes' different rates of speed, the bombers had taken off earlier and were already well on their way east. Scanning the instruments one last time, Joe saw the pistol flare fired signaling the start of the mission.

Rolling beside Capt. Adams down the grass runway, the airspeed indicator rose passed 100 mph and the Mustangs slowly lifted into the sky. Fully loaded with thirteen hundred rounds of 50-caliber ammunition and two 75-gallon auxiliary wing tanks, the Mustang now felt sluggish on the controls, but the ground soon slipped away and the airfield disappeared behind him.

A slight cloud cover obscured the ground as Adams and his wingman made the 90-degree left turn to circle and await the rest of the flight. Glancing down his wing while he turned, Joe could see the next two fighters, clearing the end of the field, M7-B visible along the side on one of the planes.

"Browning," Joe thought, wondering if Rob was as anxious about this first mission as he was. In short order, all sixteen fighters were airborne and heading toward the Channel.

Because German radio detection equipment was sensitive enough to pick up radio chatter, radio silence was the standing order and Joe could see the pilot's hand-signaling each other while they made their way toward the French coastline.

Each Mustang carried 269 gallons of fuel in the three internal fuel tanks and another 150 gallons in the external drop tanks slung beneath the wings. Routinely, pilots took off on internal fuel, in case of a sudden abort during takeoff, switching to draw fuel off the drop tanks once airborne. The newer P-51B Joe was flying, however, had an additional 85-gallon fuel tank directly behind the pilot's seat to afford an even greater range. This tank, he had learned from the other pilots; until you burned off at least 20 gallons, upset the plane's center of gravity and gave the plane a noticeable tail-heavy characteristic which could cause the nose to pitch up suddenly if too much back pressure was applied to the control stick. Not wanting to deal with that while trying to maneuver in a dogfight, Joe left the fuel selector set to this internal tank for a while longer.

Keeping one eye on his flight leader's wing, Joe scanned the controls and checked his oxygen system, climbing through 10,000 feet. At 16,000 feet, the Merlin's steady drone changed pitch with a noticeable *clunk* as the Merlin's supercharger automatically came on maintaining the engine's performance in the thinner air. After 20 minutes, now at 28,000 feet, the fighters caught up with the bombers flying 5,000 feet below.

The bombers had been easy to spot, their course clearly marked by the long white vapor trails stretching out for miles behind each engine.

"No chance of sneaking up on the target today" Joe thought while he moved the fuel selector to draw from the external drop tanks. It was reassuring how automatic his actions were in operating the plane.

"So far, so good" thought Joe, taking a deep breath and feeling the cold hard rubber of the oxygen mask against his face. "Now all I have to do is not screw up when the shooting starts."

Looking over once again, Joe could see Capt. Adams, his Flight Leader, slightly in front of his port wingtip. On the other side, Joe noticed the two other Mustangs making up the rest of Blue Flight.

Lt. Perkins flying *"Daisy Mae,"* M7-P with Lt. Browning in M7-B flying his wing. Somehow knowing Browning was there helped ease the tension. The image of Browning lying flat out on the floor of the briefing room was still fresh in his mind.

By this time the Mustangs had caught up with the Havocs, they had tested their machine guns and were flying a zigzag pattern so as not to out run the slower bombers. Continuing to scan the sky, Joe noticed Lt. Perkins rocking his wings to alert his wingman. Something was wrong and he was signaling for an abort. Looking closer, Joe noticed a thin white stream of liquid coming from beneath Perkin's cowling. The Lieutenant gestured something to his wingman and dropping out of formation, headed back to the base. Apparently he had instructed Browning to double up with Adams who now had two wingmen, Joe and Rob, both rookies.

"I hope Browning stays off my tail." Joe thought, recalling their close encounter on the field earlier.

Up ahead, Joe now saw the approaching French coastline. Their flight route had taken them over toward the Belgian border in order to avoid the heavy ack-ack and fighters around Abbeville and simultaneously hoping to confuse German radar about the intended target. Twenty-five minutes later, while passing over Cambia, Joe saw his first *Fliegerabwehrkanone* or "flak," the German anti-aircraft fire.

Small black puffs of smoke appeared on the horizon directly in front of the bomber formation with only an occasional puff at the higher

## My Shadow

altitude of the fighters. None of the other fighters seemed to be paying it much attention so Joe tried to do the same, with far less success.

Moments later, off to his left, came a blinding flash that shook Joe's plane so violently he thought he'd collided with Captain Adams. Looking over, all he saw was Rob's plane fighting to recover as well. Blue Flight leader's plane was gone. Captain Adams was gone!

"Flak!" The thought flashed through Joe's mind. "A direct hit!" The captain, his plane his Flight Leader, were gone! At that same instant, and without thinking, Joe pulled hard right, dropping his wing and rolling out of formation dodging imaginary flak.

Leveling off 1,000 feet below, Joe saw the bombers, now much closer, starting their bomb run. Above, he could still see the rest of the fighter group holding position.

"Son of a bitch!" Joe chastised himself for breaking formation. The sudden disappearance of his Flight Leader had shaken him badly and in a panic he'd thrown his plane all over the sky.

Taking a deep breath to regain his composure, Joe began his climb back up to the rest of the group. While doing so, he glanced into his mirror and was surprised to see another Mustang right on his tail. As he continued to climb, M7-B pulled up along side him and Rob, looking confused, made a gesture that clearly meant, "What just happened?"

Joe shook his head and signaled for the two to rejoin the group.

Forming back up with the rest of the fighters, the bombers, having finished their bomb run, began their evasive turn to start the flight back across the Channel. The Havocs had done their part and from what Joe could see, the repair base was a smoking shambles. Then someone broke radio silence.

"Bandits... bandits... bandits, eleven o'clock high!"

Joe looked up and to his left, spotting the enemy fighters and then back over at Rob on his wing as if to say, "What now?"

"Blind leading the blind" came Rob's reply over the radio.

"Cut the chatter and release drop tanks," came a curt order from a voice Joe didn't recognize, but he followed the order just the same.

Like a flock of eagles, the 14 remaining Mustangs all released their drop tanks at once and turned hard into the oncoming fighters that bore down into the group of Havocs. Joe could see small flashes along the wing edges and cowlings of the enemy planes that indicated deadly machine gun and cannon fire.

Certain that he was about to be either shot down or collide with one of the Germans, Joe followed the rest of the squadron as the enemy planes raced passed in a blur. Despite the converging speed, well in excess of 600 mph, Joe could clearly see the black crosses painted on the light blue side fuselage of each Focke-Wulf 190, *Würger* or Butcher Bird as they passed and...

"...God help us, they've got yellow cowlings! It's the Abbeville Boys!" Joe shouted yanking the control stick hard into his gut. "This is one hell of a baptism by fire!"

Despite the external temperature of -40 degrees, sweat ran down the inside of Joe's flight suit, his breath coming in short rapid gasps. Pulling the Mustang into a sharp bank, the scene around him lost color and started to dim as the g-force of the hard, climbing turn drained the blood from Joe's brain, his arms feeling like lead.

"I will NOT pass out," Joe intoned with conviction as he tightened his leg muscles and rolled out at the top of his climb. Color returned to his vision and looking up, there, right in front of him was an FW-190 rolling left and diving on one of the bombers.

Pausing for no more than a heartbeat, Joe quickly ran through his mental checklist:

"Drop tanks? *Gone!*
Gun sight? *On!*
Guns? *Charged and tested."*

Pushing slightly forward on the stick, Joe placed the pipper of his N-3B gun sight on the body of the plane before him. Steadying the plane for just an instant, Joe pulled the trigger on the stick and feeling the plane shake, four steams of bullets shot out before him passing just above and to the left of the 190's cockpit.

"Shit! Missed him!" Joe said, disgusted with his poor aim, but the 190 broke off pursuit of the Havoc, rolling hard right. Joe adjusted his controls to pursue the escaping Focke-Wulf.

The radio chatter from the other fighters had now become a cacophony of excited voices all describing the current battle, but somehow Joe clearly made out the words:

"Joe, break left, NOW!"

Without so much as a thought, Joe threw his fighter over watching streaks of white light the size of baseballs come flying passed his cockpit. In an instant, he knew exactly what he was seeing, deadly tracer rounds aimed right at him! In all the confusion, he had violated the first rule of aerial combat; always check your tail; the error had almost cost him dearly.

Pulling up hard, Joe doubled back around in time to see a Mustang closely followed by a 190 and diving away hard.

"Come on!" Joe again heard in his headset, "He's all set up for you. Take the shot!"

This time he recognized the voice. It was Browning's, and with a Focke-Wulf on his tail and firing. Looking more closely, Joe could also

## My Shadow

see the *Gruppenkommandeur* chevrons on the fuselage and multiple victory markings on the plane's rudder. This was no rookie pilot!

Whipping his plane over hard, Joe pushed the throttle forward and rolled out just behind the two planes. Hardly believing his luck, Rob's maneuvers had placed the 190 into a perfect firing position for Joe's four 50-caliber guns.

"Don't just sit there, let 'em have it!" came Rob's voice again and with that, Joe pulled the trigger. Careful not to hit Rob's plane, Joe gave the 190 a long burst of fire.

White puffs appeared all around the wing roots of the 190 and pieces of plane began flying off. Soon, black smoke was pouring out from beneath the plane as the canopy flew off and the pilot, rolling the plane over onto its back, bailed out.

"Yahoo!" came a scream in Joe's earphones as he watched the German pilot, in apparent slow motion soar past his wingtip.

"You got him! Good shooting, Georgia!"

Joe let out a long breath and wiped the sweat from his brow with the back of his glove. He had not only survived his first encounter with the enemy, but had actually scored a victory against an 'Abbeville Boy' – a *Gruppenkommandeur*, a Group Commander to boot.

Panting hard and soaked with sweat, Joe turned his oxygen setting up to 100% to clear his head, and at the same time a wave of terror shot through his body. The silhouette of a plane had appeared directly above the canopy, blocking out the sun and casting its shadow across Joe's cockpit.

"Shit!" Joe swore, certain that in all the confusion he'd been bounced by another enemy fighter.

Pushing forward hard on the control stick, Joe started a negative g-dive when his earphones crackled once again.

"Whoa! Steady big guy." came Rob's calm voice. "It's only me."

"Well, what the hell do you think you're doing? You scared the life out of me!"

"Sorry, Georgia, but Lieutenant Perkins' last instructions to me before heading home was to *shadow you*, so that's what I do. Just following his orders."

"I don't think he meant it literally." Joe blurted, deciding not to disguise the anger in his voice.

"Well either way, it was a good thing I did, I thought that Yellow Nose bastard had you for sure."

"Me too." said Joe, but this time he didn't key his mic.

Looking up, Joe spotted the other 597th's Mustangs forming up, the bombers now safely out over the Channel with minimal damage. The Focke-Wulfs were gone.

## Robert Brun

The excited radio chatter of the other pilots had returned to their normal relaxed cadence as someone's voice broke in.

"All right men... Let's go home."

Hearing this the two rookie pilots fell in with the remaining twelve Mustangs and headed back across the Channel for England. Joe had never welcomed an order so much in his life.

# My Shadow

## Chapter 5
## My Shadow

*Sunday, January 16, 1944: Made it through the first mission alive! My flight leader had to abort due to engine trouble and our wing leader was hit by flak! Ended up flying wing for Lt Joe "Georgia" Dyer. Good thing too, he really pulled my fat out of the fire! - R.B.*

  Joe and the rest of the 597th Mustangs flew in loose formation, the Dover Cliffs appearing out of the Channel mist. After what he had just been through, it was a beautiful sight. While the white chalk cliffs passed beneath him, he felt the tension in his muscles suddenly relax. He'd made it through his first mission... and lived to tell!
  Fifteen minutes later, with the airfield in sight, Joe received clearance to land. He could see one of the pilots performing a *Victory Roll*, and for a brief second the idea of doing the same crossed his mind, then just as quickly passed.
  "This is no game," Joe thought, remembering Capt. Adams, and the idea suddenly disgusted him. "I saw a man killed today!"
  Wheels down and locked, flaps extended, engine throttled back, Joe flared out the Mustang and felt the two main wheels firmly touch down on the grass field. Opening the canopy, cold damp air washed over him, and taking a deep breath the rookie pilot taxied over to the hardstand where the ground crews were waiting.
  Almost before he'd stopped, his crew chief, Zeke was up on the wing helping to unfasten the safety harness.
  "Good to see you made it back." Zeke said dryly. "I've been worried sick."
  "About me or your plane?" Joe shot back, questioning the mechanic's concern for his well being.
  "Well, my plane of course, but I'm real glad to see you didn't get any blood on her either." Zeke said using the feminine reference he did

with all his planes. "In fact," and with a quick look at the plane, "I don't see a mark on her. Quiet trip?"

"Hardly," Joe admitted. "Had a run in with the *Yellow Noses*."

"No shit?" Zeke asked in disbelief. "Anybody hurt?"

"I don't know, I don't think so, but Capt. Adams is dead," said Joe, his voice trailing off. "Flak... Direct hit."

Zeke was suddenly quiet while he finished unfastening Joe's harness. "Sorry to hear that. Adams was a good man."

In reply, and after a pause, "Better get this bird cleaned up." was all Joe said sliding down off the wing.

Walking across the hardstand, Joe watched as M7-B pulled up alongside and Tom, Rob's crew chief, ran to help Lt. Browning out of the plane. Joe slung his parachute over his shoulder and removing his flight helmet, strode over to Rob's plane. As he approached, he noticed a line of bullet holes across the left side and rudder.

"Don't you worry none, Lieutenant," Joe heard the crew chief say. "We'll have her patched up good as new by morn-in."

Lt. Robert Browning climbed out onto the wing and began removing his parachute and flight jacket. His shirt, soaked through with sweat, belied Rob's cool demeanor.

"There you are!" He said, noticing Joe standing below, "Nice shootin', Georgia."

"Thanks for covering me today."

"No need to say what lurks in the hearts of men," came Rob's reply paraphrasing the quote from the popular radio series and doing his best Lamont Cranston impression. Jumping down from the wing of his plane, Rob slapped Joe on the back, "The Shadow knows."

### Friday, April 6, 1945

Joe lit another cigarette as he continued to stare off into the darkness. His dry throat burned and his stomach ached, he hadn't eaten since breakfast over fourteen hours ago.

As if by magic, Zeke appeared with a sandwich wrapped in greasy brown paper and a thermos of coffee.

"Thought you might like something to eat, Cap?" said Zeke holding the bundle out to the pilot.

"Thanks, Zeke." Joe replied.

Taking the offering gratefully, Joe took a bite of the stale sandwich and washed it down with the lukewarm coffee. He couldn't recall when SPAM had ever tasted so good, and the tepid coffee took some of the sting out of his throat. His crew chief turned and Joe stopped him.

"Zeke?" Joe called as the man started to leave. "How long have you been over here?"

## My Shadow

Zeke turned, eyeing the Captain not sure what he was getting at, and then replied.

"Let's see... It'll be going on three years this May. Seems like most of my life now looking back. I don't think I can remember a time when I wasn't up to my elbows in fighters." Zeke replied.

Joe thought about this for a minute.

"Seen a lot since then, I'll bet."

"Cap, you have no idea," then after a pause "but stranger things have happened than fighters returning three hours overdue." There was a pause. "Now if you'll excuse me, I got a sick Merlin to look at. Captain." and Zeke, saluting Joe for the first time he could remember, turned and headed back across the field.

"Thanks Zeke." Joe said to no one in particular.

Joe thought about Rob. "You're out there somewhere, buddy" trying hard to convince himself, "and I know you're okay."

### Monday, January 17, 1944

The winter of 1944-45 was the coldest and wettest on record so it came as no surprise that Joe's second mission was scrubbed. Still feeling his way around, Joe took the opportunity to reflect on events of the preceding day. He didn't feel that he'd properly thanked Rob for looking after him during the mission, and decided to seek him out in order to do so.

Over at the repair bay, Zeke told him that he'd seen Lt. Browning over by his plane on the hardstand.

Joe headed to where M7-B was parked and saw Rob and his crew chief Tom, having an animated discussion about something. Coming around the front of the plane, Joe could see the two men looking at drawings in Rob's notebook.

"I want it as matte a black as you can make it with a big yellow sun and below in grey letters I want it to read, *Shadow."*

"What are you guys up to?" Joe interrupted, rounding the front of the plane.

"Picking out my nose art." Rob explained. "You gave me the idea for a name. *Shadow.* What da ya think?"

Joe thought a minute then glanced at the notebook Rob held out for him to see. Drawn out skillfully on the open pages was an exact replica of the front end of the Mustang with a design for the name *Shadow* shown in place on the nose. Next to the drawing was a detailed diagram showing the dimensions and colors to be used for the name.

"That's great!" Joe said, surprised at Rob's handiwork.

"A thing of beauty is a joy forever," was Rob's reply, quoting Keats. "Do you think you can get it done by our next mission?" Rob asked Tom. "I'd like to show it to the *Brats* as soon as possible."

## Robert Brun

"I'll do what I can, Lieutenant." came the reply. "Right after we finish patching up a few holes first."

"Perfect." Rob said, and turning to face Joe, he asked, "You pick out a name yet?"

"Uh... not yet." said Joe slightly embarrassed, not having given it a thought.

"Well you better get on it. Personalized planes are luckier you know." Rob recited punching Joe in the arm, and shuffled away singing *'Me and my sha-dow... strolling down the A-ven-ue.'*

"He's really something." Joe said to Tom as Rob tap-danced away.

"Seems to think the same about you," said Tom. "All he's done since he landed last night is tell everyone who'd listen about how you shot down that 190 that had him cold."

"Really?" Somewhat surprised and glancing over at Lt. Browning, dancing his way into the mess hut. Joe turned and headed back to his quarters. Perhaps he had misjudged his fellow rookie.

For the next three days, the weather remained the same, heavy low cloud cover across England and the Continent, with drizzle, occasional snow and a ceiling less than 1,000 feet. It was miserable. Cold and damp, and for a fighter pilot not flying, it was torturous.

Joe had been restless all morning. He had little interest in the book he'd borrowed from Lt. Taylor with whom he shared his quarters and had already written two letters to his girl back home. After a while, he started to think about what Rob had said about naming his plane. Giving it some thought, he knew it would have to be something special, something that suited his personality. There were plenty of personalized planes in the squadron already, names like:

*Bank's Bandit*
*Butcher Boy*
*Read 'em and Weep*
*Taylor's Maid*
*Hell on Wings*
*Rosie's Rivets* and
*Dina-Might* just to mention a few.

There were also many named for sweethearts back home, and the occasional, anatomically enhanced pin-up.

None of these felt right to Joe however. He wasn't the type to brag or bluster about being a fighter pilot, and he certainly wasn't going to put a half-naked painting of his girl Maureen 'Mo' as he called her, on the fuselage for all to see, no matter how much he missed her! Thinking for a while longer, he had an inspiration.

## My Shadow

Although against regulations, he, like many of the pilots, had always carried Mo's picture with him when he had flown during training. It was the silly one she'd given him on his first leave back home after getting his pilot's wings. In the photo, she had her arms extended like wings pretending to fly and had written, *"I'll always be flying with you,"* at the bottom. And so she was.

"Regulations be dammed! *Mo & Joe* it is!" he said to himself smiling, and wrote it down on the notepad there by his cot, to see how it would look.

No, something wasn't right. Looking at it for a minute or two, he picked up the pencil and crossed out the *e* and the *&* replacing it with a hyphen. *Mo-Jo*, it now read. He wrote it out again, *Mo-Jo*.

After half an hour and three more sheets of paper, Joe had written the name out just about as many ways as he could come up with. Finally, out of frustration he scrawled *Mo-Jo* in his own handwriting. That was it! It looked perfect.

Joe placed two sheets of paper against the hut window and carefully traced the name onto the new sheet of paper. Almost giddy with excitement, he stuffed the drawing into his coat and heading out into the pouring rain, ran across to Rob's hut and in through the door with a bang. As he had expected, Rob was sitting knees up on his cot working away in his notebook.

"Jeez you're wet." Rob said looking up from his writing and stating the obvious, "What's all the excitement, Germany surrender?"

"A... name..." Joe said, out of breath and holding out the sheet of paper, "I picked out... a name for my... plane."

"Great. Let's see," Rob said, getting up and walking over.

"Mo-Jo," Rob read aloud, and for a moment Joe was afraid Rob disapproved of his choice.

"Mo, uh, Maureen, my girl back home and me... Her friends call her 'Mo'. Mo, Jo... Mo-Jo? No `e.'" and as he said this he showed Rob the photo of Maureen.

Rob took it and let a long whistle gazing at the photo Joe had given him. "She's an angel, and with wings to boot."

"Yeah?" Joe said taking the photo back from Rob's hand, not quite sure he liked the way Rob had been looking at it.

"Mo-Jo..." Rob said aloud and paused for a moment, "I like it, in fact... I love it! Sounds real *"jive!"* as the jazz musicians back in *Bald-di-more* would say. Just one thing," and with that Rob snatched the drawing from Joe's hands and took it over to a small desk in the corner of the room.

Picking up the pencil from his cot, Rob turned it over to the eraser and rubbed out the hyphen between *Mo* and *Jo*. Then, placing the page

55

along the edge of the desk tore the sheet in two, right between the names and slid the *Jo* down and just a bit closer to the *Mo*.

"There" he said stepping back, "I think that's got it."

Joe wasn't sure he approved of what had just happened, having his drawing torn in half, but as he stepped forward to look at the damaged drawing, he had to admit, it did look better.

"I hope you don't mind my making a few adjustments, buddy?" Rob asked, almost apologizing while he taped the pieces back together. "It's kind of a bad habit I've developed when I see *near greatness.*"

"No... no... I think you're right," Joe said, once again excited, "It's better. You made it better. Thanks."

"Don't mention it - besides, it was my pleasure."

# My Shadow

## Chapter 6
## Fighters Over France

*Wednesday, January 26, 1944: Picked out a name for my plane. "Shadow." Georgia gave me the idea and it seems to suit me. Considered "Paula," but I don't think I could handle those questions just yet. He named his plane "MoJo" after him and his girl back home. Seems really hung up on her. Must be nice - R.B.*

Two days later when the weather finally cleared, the 597th Fighter Squadron's *MoJo* and *Shadow*, now sporting their distinctive nose art, were once again out on the flight line awaiting the signal for take off.

With Rob's guidance (and twenty bucks), Tom had done an exceptional job painting the name on Joe's plane. He could almost feel Maureen's presence there with him in the cockpit. Even the plane's Merlin engine seemed to run smoother, and although Joe knew that was just in his head, he liked the feeling all the same.

He'd written Maureen the day before with the news and had even gotten one of the crew over at the propeller shop to take a *color* photo of him standing by the nose of the plane sporting the name. He hadn't felt this excited since he was ten years old and had used his mother's nail polish to paint *Red Devil* on his first two-wheeled bike. He could still recall the spanking his father had given him for it too – but it had been worth it!

The mission this day was to escort forty B-24 Liberators over to France, the target, a munitions factory just outside of Paris. Joe had been assigned to fly wing to Capt. Jim Gauthier, one of the top veterans of the 597th and a seasoned pilot with 52 missions under his belt.

Gauthier was a stocky, muscular man with a shock of pure white hair, despite his young age of only 28. He had been flying with the 597th

## Robert Brun

since the group was first formed, and had shot down six confirmed and three probables. Joe knew he could learn a lot from flying with this ace.

That morning at the briefing, Weather Command had reported a thick cloud cover from 3,000 to 8,000 feet over all of East Anglia. This low front, they were told, would begin to break up over the English Channel and finally clear once over the European continent. Groans were heard from the other pilots who knew well this would mean a close formation climb through the first 5,000 feet of almost zero visibility.

"Hey, Joe!" Rob called approaching the flight line, "I see Zeke's stenciled your first victory cross on *MoJo*. Way to go buddy."

"Thanks, but if you hadn't been there, I'd probably be just another mark on some Hun's rudder. Maybe you'll get your first today?"

"Hey, don't mention it. Just did what I was told, looking out for you. That was some fine shootin' though I'll tell ya. Anyway, you do what you want with your plane, as for me, I don't want any of those stinkin Nazi swastikas marking up my *Shadow.*" Then, after a pause, Rob asked, *"Who you flying with today?"*

"Gauthier," Joe replied, still a bit puzzled by the note of hesitancy in Rob's remark.

"Great! I'll be shadowing Captain Rowan today. I've heard good things about Gauthier. I understand he's got six confirmed?"

"That's what I hear. There's a lot I can learn from a guy like that."

"Well, good hunting, Georgia. See you upstairs." and Rob headed over to his plane.

Joe made his way over to *MoJo* where his crew chief was busily finishing up preparations for the mission.

"Zeke, what do you make of Browning? Joe asked handing his parachute to the crew chief.

"Beats me, I can't figure the guy. I mean, he seems like a competent pilot, but a little outta trim if you know what I mean. Always seems to have his nose buried in that book of his, too."

"I know; it's funny."

Assisting Joe into his chute, Zeke cinched up the safety straps and Joe settled himself into the cockpit.

"She'll be reading a bit high on the oil-pressure gauge, but that's nothing to worry about. Needs the filter changed is all. I'll have it done before the next mission. Anyhow, take care of *my* plane Lieutenant?"

"Will do, *Sergeant*." Joe quipped with a mock salute while Zeke closed and locked the canopy.

Joe started the Mustang on Zeke's signal and, after the required warm up, he instructed the ground crew to remove the wheel chocks and taxied *MoJo* out of the hardstand.

## My Shadow

Taking off again in groups of four, Joe pulled in tight on Capt. Gauthier's tail and the two planes climbed away from the field. Looking up ahead, he could see the solid light-grey cloud ceiling above him. At 3,000 feet, the clouds engulfed the entire flight.

It took all of Joe's training, skill and concentration to keep the faint silhouette of the Captain's plane in view. Eyes darting back and forth from the artificial horizon to the Captain's wing Joe was soon surrounded by clouds so thick he could no longer see the tip of his own wings.

Several times, he thought he'd lost the Captain altogether, but before pulling further away; as was standard flight procedure in such conditions, he'd just make out Gauthier's wingtip again and eased his plane back into position.

Flying in these conditions was hard on the nerves and mid-air collisions became a very real possibility. On several occasions during the climb, Joe actually felt the turbulence of prop-wash buffeting his plane and knew then he was too close to the lead plane. Knowing which way to turn to correct this was always guesswork, but today, Joe was lucky, he'd guessed correctly and the ghostly specter of Gauthier's plane again appeared before him.

After what seemed like an eternity, the sky began to brighten, then instantly, and without warning, Joe broke clear of the clouds, finding himself right off the wing of his flight leader's plane.

The sudden brightness was blinding, but it was with a feeling of great relief that Joe pulled the tinted goggles down to shield his eyes while viewing the scene around him.

As far as he could see, everything had the appearance of white cotton wool beneath a deep blue, cloudless sky. Like a fresh blanket of snow over rolling hills, the clouds spread out to the horizon beneath them while the planes of Blue Flight continued to climb to their assigned altitude of 28,000 feet.

Looking off to his left, Joe watched first one, then two, then the remainder of the squadron break out of the clouds and into the gleaming sunlight. Soon, all eighteen planes were above the cloud tops. Two additional Mustangs, sent along as spares in case of aborts, rocked their wings in salute and returned to base while the rest of the 597ths Mustangs reformed into their flights and headed out across the Channel in search of the bombers, the Big Friends, the B-24 Liberators.

Joined by several other P-51 groups, the fighters were nearing the French coastline when Joe first spotted the formation of B-24s skimming through the broken cloud cover below and off to his right. Heading over and dropping to 25,000 feet, he and the 15 other Mustangs of the 597th cut their speed and began flying their zigzag patterns.

## Robert Brun

To his left, Joe could see a squadron of British Spitfire Mk Vbs turn away and head for home, rocking their the distinctive elliptical wings in salute as they did so.

The Spitfire, along with the Hurricane and the men who'd flown them, were the unquestioned heroes of the Battle of Britain, handing Hitler's his first defeat in the summer and fall of 1940. Up until then, the *Luftwaffe* had faced only disorganized air forces flying obsolete aircraft in the hands of inexperienced pilots. German *Reichsmarschall* Hermann Göring assured the *Führer* that his mighty *Luftwaffe* could easily sweep aside the small, overtaxed Royal Air Force, just as they had done with the pilots of the French *Arme'e de l'Air* and the British RAF Expeditionary Force in France. At this point, with France fallen and the United States not yet involved, Britain stood alone against the Nazi threat.

As a result of Göring's boisterous overconfidence, poor planning and the concerted efforts of the RAF pilots, (Prime Minister Winston Churchill's now famous "few") the German air force received its first bloody nose. RAF Air Chief Marshal, Sir Hugh Dowding and his "Chicks" had sent the Germans packing, but only just. It had been a *very* close fight.

Although a capable dog-fighter, the Spitfire, designed by Reginald Mitchell in 1936, was strictly speaking, a defensive fighter and lacked the range necessary to escort the bombers much farther than the French coastline. That's where the 597th and their Mustangs took over. The Mustang, with an increased fuel capacity, laminar flow wing and extended range could make the trip with the bombers all the way to these French targets and back.

The interesting thing was that the Spitfire and the Mustang were, in a manner of speaking, first cousins. British test pilot Ron Harker who, disappointed with the poor high altitude performance of the early Allison-engine powered P-51A, suggested the idea of matching this new airframe with the famous Rolls-Royce/Merlin engine. And Baron Max Aitken, a.k.a. Lord Beaverbrook, England's Minister of Aircraft Production, through backdoor negotiations, had licensed the American Packard Automobile Company to mass-produce the Packard/Merlin engines in the quantities needed.

Joe watched the Spitfires depart and settled in on Gauthier's wing while keeping his eyes open for enemy fighters. Scanning the sky from horizon to horizon, Joe kept track of the rest of the 597$^{th}$'s Mustangs who now joined the other fighter groups among the bombers. Looking off to his left, Joe spotted M7-B/*Shadow* hanging on tightly to Capt. Rowan's port wing. Joe wondered if he'd do as well covering Gauthier as Rob had

## My Shadow

done covering him on their first mission and winced as his thoughts returned to the loss of Captain Adams.

The clouds thinned and the patchwork fields of the French countryside slid past beneath the planes. The group flew on with only the occasional burst of flak to break the monotony, indicating that they were nearing the target. Watching from above, Joe switched his radio over to C-channel to monitor the bombers' frequency. The Big Friends were usually the first to spot incoming *bogies* and Joe didn't have long to wait.

"Blue flight, this is Blue Leader, we've got bandits coming in at two o'clock high." Joe could hear the audible chatter of the bomber's fifty-caliber machine guns firing. "I hope the Little Friends are listening in."

Looking down on the bombers, Joe saw small dark specks buzzing around the lead B-24 and knew each one was a German fighter.

Switching back to the fighter squadron's channel, Gauthier's relaxed voice gave the order to drop wing tanks and break into the oncoming fray.

Joe swallowed hard while he held on tightly to his flight leader's wing. In front of the bombers, German fighters spread across the sky, clustered in groups of four planes each.

This loose finger four formation the *Luftwaffe* had developed and perfected during the Spanish Civil War made maximum use of their fighters, allowing the leader to keep a lookout ahead while freeing up his wingmen to scan the skies around them. This technique along with the *head-on attack,* that exploited the bombers largely undefended nose, had served them well against the Allied bombers earlier in the year.

"Dyer, cover me, and stick close," came Gauthier's voice through Joe's headset.

Joe clicked the mic in acknowledgment and pushed the nose of his plane over to follow.

As they approached the bomber stream, Joe watched the *Tail-End Charlie*, the last bomber in the stream, being viciously attacked, battling with at least four 109s.

"Let's see if we can break up this party." Gauthier barked as the two Mustangs dove down into the fight.

Following his flight leader, Joe kept alert for enemy planes while Gauthier opened up on one of the 109s harassing the *Charlie*.

*Tail-end Charlie* was the position assigned to the newest crew to arrive with a bomb group. These crews were green, nervous, and to make matters worse, also had to continually deal with the propeller turbulence of all the other bombers ahead of them. This made for a rough ride and a lot of work for the new pilot to stay in formation. Paying their dues.

Joe empathized with the *Charlie* crew, this being only his second mission, until, on getting closer, a line of white-hot tracers shot out from

## Robert Brun

the Liberator's rear turret passing uncomfortably close to his canopy! These guys were so *flak happy* they were shooting at anything with a single engine, which at this moment were Joe and his flight leader.

Seemingly unfazed by the friendly fire, Gauthier opened fire and blowing the wing off one of the attacking 109s, sent it spinning out of control into the thinning clouds below. The Messerschmitt's two remaining companions then split up, breaking left and right.

"I've got the left one, you take the right," crackled Gauthier's voice over the radio.

"Roger," Joe replied. He felt a bit uneasy leaving his flight leader, but followed Gauthier's order. Chasing after the second 109 as it dove, Joe was now flying almost straight down.

The P-51's canopy shuttered and the airspeed indicator climbed into the red as Joe gained on the fleeing Messerschmitt. He squeezed off a few short bursts to check his aim when, without warning, the 109's pilot yawed his fighter hard to one side rapidly loosing speed. Before Joe knew what was happening, he shot past the German fighter, barely missing the plane. In less than two seconds, Joe had gone from being the hunter to being the prey.

Craning his neck around in a panic, he tried to spot the 109 and pulling hard into a steep, climbing turn, Joe found himself again fighting to keep from blacking out.

His arms heavy from the strong g-force of the maneuver, Joe advanced the throttle lever to full power and *MoJo* clawed for altitude. Breathing rapidly, Joe's vision returned, but the 109 was nowhere in sight.

Still unsure as to the whereabouts of the enemy fighter, Joe jinxed his plane all over the sky, afraid to fly straight and level for more than a few seconds. Before long, his fears were confirmed when 20mm cannon fire shot past his canopy. Glancing into his rear-view mirror, Joe found the 109 he'd been looking for, on his tail and closing fast.

Breaking right, Joe rolled *MoJo* onto its back and pulled for the ground gaining speed and hoping to loose the pursuing German. It was no good, the Jerry pilot stuck to his tail like glue, firing continually as he did.

Joe rolled out of the dive and skidded his fighter right, then left then right again before leveling off as he shoved the throttle lever through the *gate* that activated the Merlin's emergency water injection system.

With this sudden burst of power, Joe began to pull away from the enemy fighter all the while seeing more tracer fire, but this time of a smaller caliber. Seconds later, another Mustang shot past, its air scoop just clearing Joe's spinning propeller disc, followed by the sound of hysterical laughter in his earphones.

"Bra-ha-ha-ha-ha! The Sha-dow knows."

## My Shadow

Flying just above his canopy, M7-B/*Shadow* rolled up and away without further comment. The pursuing 109 was gone.

Pulling up into a slow climb, Joe spotted Gauthier's plane and rejoining his flight leader, the two planes climbed back up above the bombers for another look around.

"Yellow Flight leader to Yellow Flight. The Big Friends have reached the IP, let's keep sharp and pick em up once they've pickled the target."

The IP was, for Joe, the worst part of an escort mission. In order to place the bombs accurately on the target, the bombers needed to fly straight and level while the bombardier sighted the target. During this part of the mission, it was the bombardier, with the help of the top-secret Norden bombsight who actually flew the plane. This, however, also made the bombers perfect targets for the German 88 and 105 mm anti-aircraft gunners, and these gunners fired with deadly accuracy.

The flak was thick over the munitions factory and Joe saw two bombers hit and start down, twenty men. How many chutes?

Low on fuel, the German fighters had vanished for the moment, but Joe knew they'd return soon enough.

Joe and Captain Gauthier rejoined the rest of the escort fighters while the bombers completed their run before breaking off to escape the flak. Circling around in a wide turn, the B-24s formed into the defensive boxes so they could cover each other during the return flight home.

Flying top cover, Joe prepared himself for another wave of enemy fighters he knew would come. Unhappy with his performance thus far Joe resolved to keep more alert. When the *Luftwaffe* did return, he'd be ready.

# Robert Brun

## Chapter 7
## Fighters Over France part II
*Wednesday, January 26, 1944*

Joe and Captain Gauthier had no sooner caught up with the rest of the Liberators, then a second wave of enemy fighters approached in the distance. Two of the B-24s, suffering battle damage, had fired off the twin green flares signaling a request for a fighter escort out of trouble. With multiple engines knocked out and propellers feathered, the two bombers were unable to keep up with the rest of the formation. Stragglers were easy meat for the *Luftwaffe* fighters.

A decision had to be made whether to continue with the bomber formation or drop back and protect these stragglers. Fortunately, this was a decision Joe didn't have to make. Gauthier's steady voice broke in over the radio.

"Yellow two, this is Yellow Flight leader. Do you read? Over"

"Yellow Flight leader, this is Yellow two. Over."

"Dyer, we got a few stragglers with battle damage trailing behind the flight. These guys don't have an ice cube's chance in hell when those 109s arrive. I'll signal the rest of Yellow Flight to keep close to the Big Friends while you and I hang back here and see if we can keep these guys outta trouble until we reach the coast."

Acknowledging the order, Joe then switched over to the bomber frequency and listened in while Gauthier relayed the message to the two damaged B-24s. Only one of the two replied, Joe assuming the other's radio had been knocked out. He could hear both the relief and fear in the bomber pilot's voice as he acknowledged the call.

## My Shadow

Switching back to the fighter channel, Gauthier instructed Joe to take up a position above and behind the two Liberators.

Flying top cover, Joe surveyed the damage to both planes. The tail turret of the right bomber had been completely shot away. The body of the gunner lay slumped over as a fellow crewman tried to help him. A hole the size of a card table had been blown out of the main fuselage, and black smoke streamed from the #3 engine.

The second bomber seemed in worse shape. Both starboard engines were feathered causing a noticeable yaw that pulled the bomber unevenly to the right. The Liberator's twin rudders at full deflection could barely compensate for the planes yaw resulting in the bomber crabbing along at an odd angle. It was a wonder it was flying at all.

In order to stay with the damaged bombers, Joe flew just above the Mustangs stall speed. Even retarding throttle as much as he dared and adjusting the pitch of his prop, Gauthier and Joe could not keep from overrunning the damaged planes. The two fighters had to weave too far out to the side to adequately cover their charges.

"Yellow two," Gauthier called, "lower flaps to 20 degrees and maintain position as long as possible."

Joe acknowledged, but knew this would mean putting both fighters into a very vulnerable position should enemy planes attack.

*MoJo* now took up a perceptible nose high attitude, feeling the stick shuddered slightly as the air speed dropped, but the two Mustangs continued to hang in with the bombers.

The four planes limped toward the French coastline when the first of the 109s returned. Before Joe or Gauthier could react, the rear B-24 was stitched with machine gun and cannon fire, its number four engine erupting into flames. The pilot had alerted the crew to bailout and soon chutes began appearing from the stricken bomber now in a long, slow, descending roll with the attacking 109 still in pursuit.

"Dyer!" came a call on the radio. "There's bound to be more where this one came from. You hang back with the remaining bomber and I'll deal with this bastard."

Joe watched Gauthier drop his nose, retract his flaps, and dive after the 109, disappearing from his sight.

Taking a quick look around, things seemed quiet for the moment. Pulling up along side the remaining bomber, Joe could see the pilot through the shattered glass of the cockpit. Noting the circle "D" markings and tail number on the crippled bomber, he retuned his radio back to C-channel, the bomber's frequency and depressed the transmit button.

"Liberator, 916, this is Little Friend Two. Do you read? Over."

"Little Friend, this is the *Georgia Peach*. We read, over."

"*Georgia Peach*. Now there's an irony." Joe thought.

# Robert Brun

"*Georgia Peach*, this is Little Friend. What's your status?"

"This is Lieutenant Saunders, Sir," came a shaky voice with a distinctive southern drawl Joe recognized immediately.

"The Captain's hurt real bad. We got two engines out and multiple wounded, but we're holding together so far. Not a very good first mission, Sir." Joe realized it was the *Tail end Charlie*.

" How many guns do you have left?" Joe asked.

"Left waist gunner is dead and I can't get an answer from the tail. Top, nose, right waist and ball turrets have all checked in, but they're low on ammo."

"Roger. Keep alert and hang in there - we're gonna get you home. What part of Georgia are you from, Lieutenant?" Joe asked, trying to calm the boy.

"Macon, Sir. The pilot's from Jefferson and our flight engineer and waist gunners are both from Savannah... er, they were."

There was a long pause and at that moment, Joe envied the camaraderie of the bomber crews. Despite the inherent and constant danger they had to endure, simply knowing that there was always someone you could count on and to share your fears with when things got bad appealed to the young fighter pilot. He suddenly felt very alone.

"Y-all from Georgia?" the Liberator pilot called back.

"Sharpsburg's my home, Lieutenant."

"No kidding?"

"No kidding. I bet you never thought you'd see France, did you?"

"Certainly not like this, Sir."

Continuing to scan the sky, Joe saw that Gauthier's suspicions had been correct. Off to his left he spotted three small dots banking hard onto an intercept course.

"We got company, Lieutenant, bandits four o'clock high. Get ready and have your men hold their fire until they've acquired their targets. Let's make every round count. I'll do what I can out here."

Joe didn't like the odds, but didn't see any point in complaining so raising his flaps he advanced the throttle and spun the Mustang around to engage the incoming fighters head-on.

The three Germans were line abreast when the four fighters converged. At 300 yards, Joe opened fire and was surprised to see the number two plane begin to smoke and drop out of formation.

Joe broke hard left while the two remaining fighters flashed passed, ignoring him and their fallen comrade, set on going after the crippled bomber.

Pulling hard around, Joe circled back to see the second 109, drawing concentrated fire from the top and right waist gun positions of the B-24, begin to smoke.

## My Shadow

Impressed with the marksmanship of the bomber's gunners, Joe took off after the remaining 109 and was interrupted by Gauthier's voice in his headset.

"Yellow two this is Yellow one. Climbing up from 3000' on reciprocal heading. What is your status? Over."

Without giving it any thought, Joe keyed the mic while closing in on the remaining 109.

"I'm a bit busy at the moment, Captain," Joe answered lining up the 109 in his gunsight and triggering his machine guns.

The Mustang shuddered as the four machine guns activated and the 109 weaved slightly before rolling over and diving for the ground.

"Number 3?" Joe wondered to himself and took another quick look around.

There above him, still limping along on two engines was *Georgia Peach*. Off to the right, Joe saw Captain Gauthier's Mustang climbing back up to join him. As far as he could see, there were no more enemy fighters and, even better the French coastline was visible ahead. What a beautiful sight! They were going to make it.

# Robert Brun

## Chapter 8
### Fighters Over France Part III
*Wednesday, January 26, 1944*

After returning to the base, Joe spent the rest of the day in his hut replaying the mission in his head. He had just lain back onto his cot when Browning entered whistling some unidentifiable tune.

"How'd it go back there today?" Rob asked sitting himself down backwards on the chair by the bed.

Joe removed the arm covering his eyes and looked over at Rob.

"Okay, I guess? Managed to bring one of the Libs home, but lost the other one to fighters. Gauthier shot that Hun down though."

"I understand you took care of a few yourself. That makes what.... three for you, so far?

"Yeah, three I think, I don't know, ask the Adjutant, he keeps those records; I just fly the planes."

"That's not what I hear. Word around the mess hall is we got a call from the 392nd Bomb Group earlier. Seems the co-pilot of the *Georgia Peach* says you took off after those three 109s like some kind of *mad dog*! Gauthier's already talked to the Colonel about putting you in for a commendation for your saving that Charlie.

"Huh. Is that so?" Joe said leaning back and placing his arm back over his eyes.

"Yep, and at the rate you're going, you'll make Ace before the end of the month."

"Hey Rob," Joe asked glancing out from under his arm, "You mind taking a powder? I got a whopper of a headache!"

"Sure thing *Mad-dog*. Oh, by the way... letter from home," and, tossing an envelope onto Joe's cot, Rob turned and left.

Joe lay there for a long time replaying the day's mission over and over in his head. He was feeling pretty good about knocking down those two 109s and escorting the one Liberator safely back home, but the disturbing image of that other B-24 going down in flames kept reappearing in his mind. Ten men, had anyone made it out? He had been far too busy to see. How many dead? How many captured? What had he overlooked and could he have done more?

## My Shadow

Joe looked at his watch. 21:45. He got up from the cot and pulled on his flight jacket. Bending over to retrieve the envelope that had fallen onto the floor, Joe noticed Mo's neat cursive handwriting on the front. Not feeling quite ready to alter his mood, he stuffed the letter into his jacket.

He walked out into the night, surprised at how calm it was. Not a breath of air moved.

Off in the distance, a warm light glowed from the window of a neighboring farmhouse and he wondered about his Mom and Mo back in Georgia. What were they doing tonight?

Looking at his watch again, he estimated the time back home to be right around dinnertime. He longed to be there, eating his Mom's biscuits, talking about milking and feed and bovine bag balm or *anything* that would get the memory of the burning Liberator out of his mind.

Walking across the base, Joe heard sounds coming from the maintenance bays and wandered in to see the ground crews hard at work getting the planes ready for the next day's mission.

Pulling back the blackout curtains, the bright glare of the fully illuminated hanger hurt his eyes and Joe squinted as he entered. A cloud of cigarette smoke wafted through the rafters while men clamored inside the open panels of the fighters, removing and replacing parts and repairing battle damage.

Joe caught a glimpse of Rob's crew chief Tom, buried up to his elbows in *Shadow's* open engine cowling. Out on the port wing, a Corporal with tin snips and a rivet gun was patching a line of bullet holes that traced across one wing.

The bustle and activity reminded Joe of a childhood fairytale, *"The Shoemaker and the Elves."* In the story, the Cobbler and his wife, after a long day of making shoes would collapse into bed exhausted, the shoes left unfinished. The next day they awaken to find all the work completed. Sitting up one night, the Cobbler was surprised to see a group of naked elves coming in to finish the work for them. Grateful, the Cobbler rewarded the elves with new sets of clothing. Joe laughed at the thought of these mechanics as naked elves and at the same time wondered if there was any way he could ever show his appreciation for their tireless work. When he turned to leave, Zeke came over to where he stood.

"Something I can help you with, Lieutenant?"

"How's it going, Sergeant?"

"Fine. Getting everything fixed up for tomorrow's mission."

"That one over there?" Joe said, gesturing toward *Shadow*. "The one with the bullet holes . . . Browning's?"

"Yeah, he catches more slugs than a target sock. Some of the crew are starting to refer to him as *Lieutenant Swiss Cheese,* but you didn't hear that from me. The guys have even got a pool going to see who'll

## Robert Brun

guess the number of rounds she takes before old *Shadow's* a total write-off. I think we're up to 167. Lt. Browning's sure keeping the crews busy, but he seems to get the job done. No kills so far, but he's getting quite a reputation as a wingman, in fact Captain Rowan was over here earlier asking if Browning's plane would be ready to fly his wing again on tomorrow's mission. "

"And...?" Joe asked.

"Why, Lieutenant!" Zeke shot back feigning insult. "You cut me to the quick! Hey, Dave?" Zeke called to the man up on *Shadow's* wing, "*Shadow* be ready by morning?"

"Lieutenant. Swiss? Count on it, Sarge."

Joe nodded in agreement, "Great."

"Don't mention it, Lieutenant. It's what we do. You keep breakin' `em and we'll keep fixin `em. Now if you'll excuse me, I've got work to do..." Zeke spat on the ground.

"Yeah. Thanks, Zeke."

Joe left the hanger and made his way over to where *MoJo* stood. The cold, night air helped clear his head. Gazing up at the starlit sky, he climbed up on *MoJo's* wing, and pulling the canopy open, settled into the cockpit. Somehow just being here made him feel better, more in control of a world gone out of control. Taking Mo's photo from his pocket, he slipped it beneath the gunsight, giving it a slight tap.

Without his parachute, Joe sat much lower in the cold metal pilot's seat, well below where anyone might see him. His breathing soon fogged over the canopy in the cold cockpit, providing a sense of pleasant solitude.

Fishing around in his jacket for Mo's letter Joe tore open the envelope, removing and gently unfolding the pages. Activating the Mustang's electrical system, Joe snapped on the flexible cockpit light and, focusing it onto the sheets of paper, he read Maureen's letter:

*My dearest Joseph,*

*I hope this letter finds you safe and well. I am fine and each day I keep you in my thoughts and prayers.*

*Your Mother sends her regards and I drop in on her often. She claims she isn't worried about you, but if she's anything like me, she is just keeping up a strong front. We both have been following the events over there on the radio and in the newspapers, and it looks like things are getting pretty rough for you and the other boys, but I know you're doing all you can to stay safe.*

*I wanted to tell you I took a job at the foundry in Atlanta? It didn't feel right not contributing to the war effort so when I heard*

## My Shadow

*about the job openings there, I applied and got hired. It's really nothing special, but I guess every little bit helps.*

*Mom says I look a fright in my work overalls, very unlady-like, and that if you could see me, you'd never have asked me out. Maybe she's right? Doug (that's my Foreman) says we girls are doing a great job. He's older (27 years) and in charge of all the "Kittens" at the plant as he calls us girls. I had a bit of trouble getting the hang of working the drill presses at first, but he gave me a lot of private overtime help and now he says I'm his 'best girl.' The other night, after work, he took the whole shift out to the U.S.O. Club in Atlanta and we all danced until I thought my feet would give out. He's really something.*

*Well, my dear, I have to get some sleep since work starts early and I need to take the 5 a.m. train into the city. Stay well and come home soon. I miss you,*

*Love,
Mo*

Joe could feel the pulse of blood in his temples as he reread the letter for the third time in a row. It was good to hear from Mo, but he couldn't help wondering who this guy *'Doug'* was. Despite his attempts to control it, Joe felt his blood pressure continue to rise. The words *'kittens, older'* and `private overtime help'* kept bouncing around in his head, and he didn't like it or how he was reacting. Then the image of the burning bomber came flooding back.

"Dammit!" Joe shouted out loud and crushed the letter in his fist, "I'm over here watching fliers get killed and she's out having a good time and who knows what else with some *four-F* named *Doug!*"

At that moment he hated the Nazis more than ever for what they'd done that had brought him here. Joe breathed heavily, the canopy fog freezing into a layer of ice, but after a long moment he managed to relax and re-read the letter again.

*"My dear Joseph...
My dear...
I miss you...
Love, Mo."*

These were not the words of some casual acquaintance and deep down inside, he guessed he knew it. Fate had separated them, and he would have to let fate take its course.

## Robert Brun

"Calm down, *Mad-dog*" Joe thought, suddenly feeling like the moniker suited him. He hated being this far away from the people he loved, but he also realized that there were people here counting on him and at the moment, far more than those he'd left behind. He had volunteered for the job, a job he was doing precisely *because* he loved them; the rest simply had to wait. As for Mo and *Doug*... well, he would just have to trust her.

Smoothing out the letter he carefully replaced it in the envelope and took a long look at Mo's photo then turned it over.

*"I'll always be flying with you,"* He read her words out loud. Joe slipped the photo back into his flight jacket, opened the canopy, and pulled himself out of the Mustang's cockpit.

"I guess you'd better get some sleep... *Mad-dog*!"

# My Shadow

## Chapter 9
## First Five

*Saturday, February 19, 1944: Completed our fifth mission today and not flying tomorrow. Sure could use a break. It's a lot of work keeping Capt. Rowan out of trouble, but so far so good. Georgia's proving to be quite the pilot if you can believe the 392nd BG - R.B.*

For the next few weeks, each new mission was much the same as the first two, escorting bombers on small raids over occupied France. The *Luftwaffe*, however, had been conspicuously absent. Initially, this was fine with Joe, but as the second week dragged on without opposition, he was becoming increasingly annoyed with an enemy that wouldn't come up and fight. Joe continued flying as Gauthier's wingman, and had even complained to the Captain about the apparent lack of opposition.

"Don't be too anxious to mix it up with the Jerries again," Gauthier said with a chuckle, "This war is far from over, and they'll come days when you'll pray for a milk run like these." Joe knew Gauthier was right, but it still frustrated him.

The following day Joe completed his fifth mission, flying wing for Capt. Gauthier. He was just leaving the debriefing hut when he heard Rob call out to him.

"He's made his five and is still alive!" chanted Rob, catching up with Joe outside the hut. "This calls for a celebration and seeing as how neither of us is on tomorrow's mission roster, let's head into town and I'll buy you a beer in celebration."

"You're on," came Joe's enthusiastic reply.

The US Army Air Corps supplied its pilots with a generous monthly liquor ration. After missions, a shot of scotch was provided to calm the nerves during debriefing, and libations were also often loudly enjoyed at the base's Officer's Club. Drinking, along with recounting the day's exploits became a daily ritual. But tonight, seeking a change of scenery, Joe and Rob procured a couple of bicycles from the motor-pool hanger and peddled the three miles into the village.

The village of Framlingham was a quiet place with little to offer to a couple of guys out on the town, but like all English villages, it did have a pub, and tonight, for the pilots of the 597th, that was enough.

Joe and Rob leaned their bicycles up against the pub's wall along with about a dozen or so others, and several jeeps. The bright glow from the windows and the pastoral countryside made it easy for Joe to believe the war was very far away. One of the 8th Air Force's bomber bases was nearby so the pub was nearly full with pilots and crews.

For the most part, bomber crews and fighter pilots got along okay, both sharing airborne duties, but you never could tell what might happen when liquor was involved so Joe and Rob entered the pub cautiously and made their way through the crowd and up to the bar.

*"Wadda be 'a-vin mates?"* asked a pleasant looking man behind the bar. He appeared to be in his early forties and it was obvious from his demeanor, that he was the pub's proprietor.

*"A cup-pul-a pints O-stout gov-ner."* Rob replied in a phony English accent that made Joe cringe. Noticing Joe's reaction, Rob glanced over and shrugged his shoulders.

The two pint glasses of stout were filled and set on top of the tap and Rob was surprised to receive a wrap on the hand as he reached for the glasses of light brown foam.

"You gotta wait for it," said a polite sergeant from the 92nd bomb group who didn't look old enough to drink. "They're a bit touchy about that around these parts."

"Thanks for the warning." Rob replied rubbing his stinging knuckles.

Five minutes later Rob cautiously took the glasses after seeking approval from the bartender who nodded.

Downing the lukewarm coffee-dark beverage, Joe turned around and, leaning against the bar, took in the roomful of fliers.

Most of these men had been flying almost non-stop for the last month over the Continent, so they were all looking for an opportunity to relax and unwind and as with most evenings, this meant beer and singing.

Captain Gauthier had already made his way over to an old, tired looking piano and was banging out a comical tune about an unlikely landing situation involving a Mustang and a scantily clad maiden that Joe had not heard before. Some of the other pilots from the 597th had gathered

## My Shadow

around singing along with the chorus. They left the verses to Gauthier who, Joe had no doubt, was making them up on the spot.

*Get her down, get her down,*
*Get her down on the ground,*
*A three-point landing's free of vice,*
*But a one-point landing's twice as nice!*
And so it went.

Joe scanned the many faces of the pilots. Each one, he supposed, had a story to tell, and Joe wondered what his story would be providing he'd live to tell it?

From his place at the bar, he spotted an attractive girl of about eighteen, bringing in a tray of clean glasses. He presumed her to be the pub owner's daughter and something about the confident way she carried herself, brushing off flirtatious comments and with the numerous stains on her work apron reminded him of Maureen, his own girl back home.

Reaching into the pocket of his tunic, Joe extracted Mo's photo and staring at it, his mind wandering back to that afternoon four years earlier...

### Tuesday, July 9, 1940

The sun beat down hot on Joe as he walked the two miles to Maureen Fowler's father's house, delivering the two gallon pails of milk Mrs. Fowler had ordered from his family's dairy. Usually Joe made the deliveries using the dairy's old pickup truck, but the tar-tape repair that had been holding the radiator hose together for the last few weeks had finally let go, and this time, it split the hose beyond mending. A new hose had been ordered, but in the meantime the Fowlers were out of milk.

Joe tried his best to disguise the true reason for offering to make the delivery, but he suspected his mother had seen right through him. She never said a word, however, standing on the porch with her arms crossed smiling and watched Joe trudge down the road carrying the two pails. His anticipation of even catching a glimpse of Mo made the two-mile walk and the weight of the milk pails hardly noticeable. After the first mile though, his arms began to ache and thinking of Mo, Joe's mind wandered back to when they'd met and how much had changed over the last few years.

Despite what all the fairytales said, it really hadn't been that hard for Joe to adjust to his father's death, nothing much had changed, really. He still did the same chores and still attended the same school. He still looked after his younger brothers, although that was becoming less of a task, as they grew older.

Joe had been running most of the operation of the dairy for as long as he could remember anyway, and the fall afternoon when his father died, hadn't changed things.

## Robert Brun

Of course Joe had to adjust to no longer seeing the man every day, and the small directives he'd issued about how to do this or that around the farm. He had to admit the man had taught him a lot, but somehow, despite the time they spent together, they had never been very close.

Early on, Joe had realized that no matter how hard he tried they never would be close. Heck, even his parents didn't seem all that close despite the three kids. *Efficient* was the word that came to mind when Joe thought about their marriage.

That his father had loved him in his own way Joe had no doubt. He had provided for him, his mom and his brothers the best he could, and better than some, but somehow Joe knew there must be more to a marriage and a family. He discovered this about a year later when he had least expected it.

Joe had been walking home from school on that early spring Friday. He was looking forward to the weekend and had plans to do some fishing, when he spotted a girl up ahead stopped in the road. On getting closer, he couldn't help noticing her pretty blonde hair reflecting the sun, the light breeze blowing a few strands across her face. Wearing an expression of concern while looking down and eyeing her bicycle, she also appeared to be stuck.

Approaching cautiously, Joe could see the hem of her dress tightly caught in the chain of her bike. Joe stopped just short of the girl and stood transfixed, feeling something inside of him jump.

"Well you gonna help me or just stand there?" She said looking up at the thin young interloper.

Joe stood motionless, unable to take his eyes off of hers. Something about her made him keep staring. There was something unreal about the experience and he was afraid if he were to look away, it would disappear.

"Hello?" she said finally breaking the silence. Would you happen to have a knife?"

"Uuuuh... yeah." Joe said as he began to reach into the pocket of his overalls, but then he snapped out of his daze.

"You're not gonna cut your pretty dress, are ya?" he said as his composure returned and he got a grip on himself.

"Well I don't know what else to do. I've been trying to get free for the last half hour without any luck.

"I can see that." Joe replied looking down at the area of the now grease-blackened hem.

She started to speak in her defense, but Joe interrupted.

## My Shadow

"Let me take a look and see what I can do." he said, putting down his books and getting down on one knee to have a better look at the problem.

"I'm Joe, by the way... Joe Dyer." Joe said, carefully examining the front chain wheel of the bike.

"Mo." he heard from above. "Uh, Maureen, Maureen Fowler, I mean. My friends call me Mo."

"Well, pleased to meet you *Mo-reen*." Joe said trying harder to be clever than he suddenly felt he should have.

Looking closely at the skirt, Joe could see that Mo had really gotten herself into a jam. The dress material had not only been caught in the chain, but through her attempts to remove it, it had also become wrapped completely around the crank arm and wound into the bearing hub as well.

"You've really got yourself stuck." Joe said trying hard to ignore the strange flush feeling he was experiencing. "Kind of like a fever." he thought to himself.

"Well I can *see* that," said Mo. "I already told you, I've been working at it for a half hour."

Joe took the pen knife out of his pocket and opening it, stuck it in the ground so it would be handy. Then he went around to the other side of the bike and lifted the rear wheel.

"Coaster brakes." Joe muttered under his breath. He had hoped that it might have been one of those fancy British bikes, the kind you could turn the crank both ways, but no such luck. He thought a minute then had another idea.

"Can you walk?" he asked Mo looking up and again into her eyes.

"Wearing a bicycle?" she said with incredulity.

"Well...no... Uh, yeah... just a few feet." Joe explained, now slightly embarrassed.

"I think so?" Mo answered after a momentary pause.

"Okay, here's what we're going to do. You walk the bike forward and I'll turn the pedals and pull the chain to the side until it jumps off the sprocket."

"Oh-kay?" Said Mo, not quite sure what Joe had in mind.

Standing up for the moment, Joe passed the handlebars to Mo and again got down on one knee placing his hand on the greasy chain just behind the front sprocket. Then, rising up to a squat, he looked up at Mo.

"Okay," he said, "you start walking with the bike, but *slow-ly.*"

Still not able to see how this was going to do anything but make matters worse, Mo began to push the bike. Glancing down, she began to giggle watching Joe waddling along next to the bike like a tall, wiry duck, all the while chewing his lower lip and concentrating on the job at hand.

## Robert Brun

They had gone maybe five feet and Mo could feel the material of her skirt pulling tighter when she heard a small metallic *clink* followed by a *whir* and a hollow *thud* then her hem came free. Joe swore softly under his breath. Looking over, Mo could see that Joe had fallen over backwards and her giggle had turned into a hearty laugh as she watched the sprawling figure scramble to his feet.

"There! And we didn't even need to cut your dress" Joe proclaimed proudly, holding the newly freed, though very greasy hem of Mo's dress up to show her that he had accomplished his mission by removing it and her from the bike. Then, he began to blush realizing in his excitement to show her the results of his handy-work he had lifted the hem of her dress up above her waist, exposing a fair amount of both legs and one thigh in the process.

"Oh!" was all Joe was able to get out before he quickly dropped the skirt and turned his reddened face away.

Mo smoothed the front of her dress and tried to compose herself.

"Thank you," she said with a wry smile, and climbing back onto the bike, careful to keep her newly freed dress away from the chain, she rode off down the road.

Joe stood there watching her fade into the distance and although he couldn't be quite sure, he thought he saw her shoulders moving. Was she laughing?

"How stupid can I possibly be?" he thought as he went back to collect his jackknife and schoolbooks. "Nice legs, though."

Rivulets of sweat ran down Joe's face and arms as he struggled with the two pails of milk he had been carrying for what now seemed like hours. His shirt was nearly soaked through when he rounded the corner of the lane and saw the Fowler's house come into view. He'd made it, but he certainly looked a mess. Self-conscious about his appearance, and smell, Joe quickly revised his plans and decided he no longer wanted to see Mo after all or more specifically, he didn't want her to see him.

Slipping around the back of the house, Joe set down the pails on the steps and rubbed his hands together to regain blood circulation. Quickly glancing from side to side and seeing no one around, he breathed a little easier and knocked on the back door.

Expecting to see Mrs. Fowler, Joe was startled when the door opened and standing before him was Mo wiping her hands on her apron, her damp, matted hair tied back in a scarf and her face covered with flour. She was a mess and she was beautiful and Joe instantly forgot about his own appearance. Mo let out a startled cry of surprise and covered her mouth when she saw Joe standing there looking more like he had swum the

## My Shadow

two miles rather than walked them. After a brief moment of awkwardness, they both started to laugh.

    Joe continued to see Mo regularly for the next year and a half. School dances, hayrides, and any other excuse he could find to spend time with her. She even proved to be a pretty good fisherman once Joe got over the fact that she usually caught more fish than he did. He still had to bait the hook for her, so he guessed that made it okay.

    The following December, they had held each other for hours after Joe had sprinted the two miles from the farm when he'd heard the news about the Japanese surprise attack. *Pearl Harbor,* a place half a world away that neither had ever heard of, had changed everything.

    The next day, President Roosevelt asked the Congress for a Declaration of War against the Empire of Japan, and three days later, Germany and Italy declared war on the U.S. The United States of America had entered into a two front, global conflict that had, by then, been raging for over two years.

    To Joe's surprise and relief, Mo hadn't even complained when he told her he was going to enlist in the Army Air Corp in hope of becoming a pilot. She knew how he felt. With tears in her eyes, Mo accompanied Joe to the bus station the morning he left for boot camp. She had hesitated when he tried to kiss her goodbye, but before he could say anything, she turned and ran back to her bike.

    Joe received an occasional letter from Mo at basic training and flight school, but she had never been much for writing. When she did write, it was a welcome relief to hear from her. *Absence makes the heart grow fonder* was how the saying went. Joe certainly knew it was true for him, but somehow, he wasn't so sure with Mo.

    When he visited her on leave after receiving his wings, she seemed distant and reserved. Even though she had given him her photo, a sign, Joe had thought, of some kind of commitment, she now seemed more hesitant than ever. When he asked her directly about it, she merely replied that she didn't know how to feel about someone who was going to be so far away and in harms way for so long. Joe had returned from his leave confused. The following day, he shipped out.

### Saturday February 12, 1944

*But a one-point land-ing's twice as niiiiiiiice!*

    The crescendo of Gauthier's song snapped Joe out of his reverie and he placed Mo's photo back in his pocket with a sigh.

    At the bar, Rob had just picked up his order for another round when he noticed Joe sitting all alone.

## Robert Brun

"Here's that beer I promised you." Rob said sliding the glass across the bar to Joe. "I think we deserve one for each mission so far!"

"I'll drink to that!" and raising his glass, Joe downed the pint in one gulp banging the glass on the table. "Two down, three to go."

"Make that five down, thirty to go." Rob said referring to the number of missions required to complete a tour of duty. The two men looked at each other for a brief moment then started to laugh.

"Barkeep!" Joe hollered, "Another round."

# My Shadow

## Chapter 10
## Back Form the Pub

*Sunday, February 20, 1944: Too late, too tired and too drunk to write... Good thing we're not flying tomorrow, eh today! - R.B.*

It was well after one a.m. when Joe and Rob finally left the pub at the request of the exhausted proprietor. Leaning against each other for support and balance, the two pilots made there way out the front door and over to where they had left their bikes. Unfortunately, where they had left two bicycles, now only one remained.

"*Oh, bloody 'ell.*" Rob said wobbling slightly as he did so. "Never could trust those bomber guys."

"Pilot or co-pilot?" Said Joe, turning to Rob.

"Huh?" Rob said, trying desperately to remember how his feet worked.

"Never mind." Joe said, grabbing the bike. "I'll fly." And lifting his right leg over the seat, he promptly fell over.

"Like hell you will." Rob barked, picking the bike up off of Joe, and offering him a hand up, "You take the front turret."

Joe rose unsteadily to his feet and made his way around to the handlebars while Rob balanced the bike between his legs. Taking out his Zippo, Rob steadied his lighter with both hands and lit the kerosene lamp attached to the front of the bike. Once the two were, more or less settled, Rob started through the pre-flight checklist.

## Robert Brun

"Throttle at half. Magnetos, closed, switches, on... Contact!" Stepping down hard on the raised pedal, the wobbly flyers took off down the road.

A heavy mist had settled over the countryside as they swerved their way back to the base along the winding road and had barely gone a mile when there was a loud *BANG!*

"I'm hit!" Joe said grasping his stomach and tumbling melodramatically off to one side, "Dammed sniper got me."

"All the way from Germany?" Rob said looking puzzled. "Some shot! I think we *av a punc-ture*."

"Oh crap, there goes my chance for a purple heart."

"You're shittin me, right?" Rob said looking right at his friend and then, after a pause, the two men began laughing hysterically.

"Yeah I think we'll get plenty more chances for that!" Joe replied barely able to get the words out.

After a few minutes, when the two had caught their breath, they started off down the road wheeling the bike between them. Not saying a word, Rob pulled out a pack of cigarettes and, lighting one, offered the pack to Joe. They walked along without speaking for a while; then Joe broke the silence.

"You ever think about dying?" Joe asked, over the sound of the ratcheting freewheel. There followed a long pause and Joe began to wonder if Rob had heard the question.

"Not anymore." Rob blew smoke out his nose, "I seen enough of it for one lifetime."

"But you don't have any kills yet?" Joe said confused and immediately wished he hadn't opened his mouth, "I mean..."

I know what you mean." Rob said, cutting him off.

There was another long pause and then Rob said,

"You know your Maureen?" Rob said

"You mean my girl Mo, Yeah?"

"Yeah well, I had someone like that once." There was another long pause while Rob collected his thoughts.

"Her name was Paula. I met her at a band concert in Baltimore one summer night a few years back. It was the 4th of July and I had gone down to the waterfront, right by Fort McHenry, like always, to watch the fireworks. That's where I first saw her, just standing there, all by herself watching the display. The way the light brightened her face every time one of those star shells went off with the Naval Academy Band playing music in the background... She was just beautiful. I swear I fell in love right then and there." Rob paused and took a long drag on his cigarette, then went on.

## My Shadow

"I guess she noticed me staring at her because she glanced over once or twice and then turned away real quick. I don't know what came over me because I'm usually real shy with girls, but I just went over and introduced myself.

"She was sort of distant at first, almost rude, but somehow I couldn't help feeling she was just being cautious, you know how pretty girls can be sometime." There was another long pause and somewhere in the mist, a horse snorted.

"It turned out she was new to the area, just moved there from St. Louis to work at the Sparrows Point Shipyard building Liberty Ships and didn't know anyone. Said she was a welder, if that don't beat all. Said her father had taught her how back on the farm in Missouri fixing tractors and such. She was the third of five kids, all girls, so it came in pretty handy.

"Well, after the grand finale I just didn't want it to end so I asked her if she wanted to get a cup of *Joe* and she agreed! Man, I felt like I'd just won the Irish Sweepstakes.

"It turned out we had a lot in common, but mostly that we both loved jazz. I'd been listening to it since I was a kid all around Baltimore. Used to drive the old man crazy, my sneaking into clubs down on Fells Point with all those Negro musicians. Jeez, you'd think their color was gonna wear off on me, the way my folks talked. But I didn't care, their sound was *"jive"* and I loved it. And as it turned out, so did Paula.

"We sat in that all-night cafe drinking coffee and talking until the sun came up and she said she had to get back to the plant for her shift. I dropped her off at the front gate only after she'd agreed to see me the next day. I practically floated back home.

"For the rest of the summer we were seeing each other pretty regular. We'd walk along down by the bay, went to Jazz Clubs on Fells Point, and what a dancer she was. We even talked about getting married, if you can imagine that. Heck, I was barely 19, living with my folks, delivering coal, and here I was already talking about getting married?"

Rob gave a snort, pulled up the collar of his jacket, and walked along collecting his thoughts.

"Well, one day, a couple weeks before Christmas, we were walking home from the movies and it started snowing. One of those early snowfalls where the flakes are big, you know, like chicken feathers. After about a block the ground was already pretty well covered with wet slush, but we didn't even care. Just a couple of dumb kids in love, I guess.

"It started snowing even harder so we made our way to the streetcar stand, and I told her to wait there while I ran across the street to buy us gum at the newsstand. By then it was really coming down so hard I could barely see her standing there at the stop. When I did look back

## Robert Brun

across, the streetcar was just rounding the corner and I guess Paula stepped out to check if it was the right one.

"She no sooner stepped off the curb to look, when a car came racing past the trolley, lost control on the slick pavement, swerved and hit her, and I mean - hard!"

Rob took a deep breath, coughed, and started again.

"God, it must have knocked her ten feet! I raced across the street and pushed aside the driver, who by now was out of the car and sobbing about how he was *sorry* and he *didn't see her*. Paula was just lying there, not moving...

"Dammit Joe, she looked just like she was asleep, but I could tell she was hurt real bad."

"Somebody must have seen what happened from one of the apartments buildings because an ambulance arrived a few minutes later. I was beside myself with worry when they strapped her onto the stretcher and loaded her into the back. I tried to go along, but because I was just her *"boyfriend,"* the medics wouldn't let me ride with her to the hospital.

" I finally convinced a cop at the scene to give me a lift, but by the time I got to the hospital and had talked the desk nurse into giving me her information, it was too late...

"...Joe, they wouldn't even let me *see* her!" Rob stopped, flicking the ash off his cigarette.

"Something inside me died with her that night. I left the hospital wandering the streets until morning and on into the following afternoon. Feeling completely lost, I walked down to the waterfront where we used to go and stared out over the harbor. I didn't know what I was going to do I just wanted the pain inside to stop.

"While I stood there at the edge of the bay, planning I don't know what, I heard a commotion coming from one of the jazz clubs nearby. This musician I knew was yelling something about the Japs bombing the U.S. fleet in Hawaii. So I wandered over and he filled me in on what had happened out at Pearl Harbor. I hadn't even heard of the place, but it changed my mind. Saved my life I guess... now that I think of it. The next day, I enlisted." Rob crushed out his smoke with the heel of his shoe.

"Yeah, that's enough death for me for one lifetime. I guess that's why I fly wing."

The two men didn't say another word until they reached the gate at the base. Before the Corporal standing guard had a chance to ask, the pilots chanted in unison, *"Lucky Strike Green has Gone To War,"* the password issued for that night.

"Better go sleep it off, Lieutenants," the sentry replied.

Rob gave a sloppy salute and, dropping the bike on the ground outside of his hut, went inside letting the door close with a bang.

# My Shadow

Joe stood there for a few minutes wondering about his friend then he headed toward his own hut.

# Robert Brun

## Chapter 11
## Big Week

*Sunday, February 20, 1944: Out with Georgia last night and had too much to drink. I don't remember much of the evening, but I think I spilled the beans about my past. I sure do miss her sometimes - R.B.*

The next day was another typical English late-winter morning, overcast damp, and colder than the day before. Joe had just rolled over in his bunk and closed his eyes again when he heard the CQ, Corporal Whiting's shrill voice.

"Lieutenant Dyer?" the Corporal's call cut through Joe's head like shrapnel.

"Wha...what?" Joe moaned, raising his pounding head, "I'm not scheduled to fly today. I already checked!"

"Well, then there's been a change sir," The CQ announced with a faint hint of pleasure. "You're definitely on my list for today's mission."

There had to be some mistake. He'd seen the mission board the day before and his name had not been on it. With a long groan, Joe pushed himself upright trying to ignore his frustration and the pounding in his head. Somehow there had been a *SNAFU* and Joe was roused out of his cot with the rest of the men. Briefing was in an hour, and feeling like death warmed over, he got up and dressed.

Although heavily overcast, the day seemed unusually bright. His eyes felt like they had been sandpapered, and his mouth tasted like an old pair of work socks. Even the familiar rumble of the distant bombers flying over was barely noticeable compared to the pounding in his head.

"*Guinness is good for you. Guinness, gives strength'.* Bah!" he thought, recalling the posters at the pub showing a man carrying a horse. "Yeah, right! No more stout for this Yank for awhile!"

## My Shadow

Entering the mess hall, the rest of the men didn't look much better than he felt. Conversation was at a minimum and few were eating, but the coffee was going fast.

After chow, Joe made his way to the briefing hut just prior to Colonel Tomlinson's arrival. He had shaken most of the cobwebs from his head and felt almost human again.

The CO entered the hut and the men came to attention. At the same moment Tomlinson reached the front of the room, Rob dashed in looking bleary and took a seat way in the back.

"All right men, listen up." Tomlinson barked across the room at a volume that racked Joe's head.

"We've received new orders, straight from Bomber HQ and we need every available pilot." Tomlinson scanned the assembled men noting their condition, then continued.

"For the passed few weeks we've been chipping away at Jerry with these smaller raids, but starting today we're going to make up for it." With that said, the assistant pulled back the curtain exposing a map of the continent showing multiple lengths of red yarn that stretched almost across the entire map.

"Men," Tomlinson continued. "Today the gloves are coming off. HQ believes that the way to end this war is to knock out Jerry's ability to make their weapons of war. That means hitting industrial centers and aircraft factories."

"A swift kick in the balls ought-a do the trick." Capt. Rowan voiced from the rear followed by a wave of laughter that helped ease the tension.

"You got that right, Lieutenant," Tomlinson's replied, "And that's exactly what we have in mind here. Today, you'll be escorting the largest bomber formation in the history of aerial warfare.

"As you know, the British have been punching back at the Germans since 1940, but only at night and with minimal strategic effect. Churchill seems to think that concentrating their raids on the cities can demoralize the people into a civilian uprising against the Nazi government, demanding an end to the war. He obviously doesn't know the Nazis." There were a few light chuckles from the men assembled.

"He seems to have forgotten that the Germans already tried the same thing during the Blitz and it only strengthened the Limey's resolve.

"The boys upstairs now think that our guys, with a series of carefully coordinated daylight attacks, can pinpoint and knock out Germany's military production facilities and strangle their ability to supply their armies in the fields.

"Unfortunately, in the last few years, because of British bombing efforts, the German's have moved much of their industrial manufacturing out of the Ruhr Valley and closer to the Capital, Berlin. This has made for much longer missions, and up to this point, the bombers have had to go in on their own, unescorted. Jerry's had a field day picking them off and losses have been too high to maintain."

Joe shuddered at his recollection of the October 14th mission - *'Black Thursday'* and the *Bloody 100$^{th}$* - the unescorted mission to knock out the ball bearing factory at Schweinfurt. The *Luftwaffe* sent up over 1,100 fighters as a welcoming committee. Sixty-five bombers lost, 650 men killed or captured. Tomlinson continued.

"Until now, that is, and that's where you men come in. The Mustangs that you've been flying have a much greater range than the P-38s and Thunderbolts. This increased range will allow the 597th to escort the bombers all the way into the target... and back!

"If all goes according to plan today, there'll be more than one thousand B-17 and B-24s airborne, carrying more than 4,000 tons of bombs right into the Third Reich's Industrial center, and you men are going to escort them all the way." Tomlinson paused to let this sink in.

"Unfortunately, we don't yet have enough Mustangs to go around so there'll still be Lightnings and Thunderbolts for part of the trip, but once they reach *Bingo* you'll be on your own."

A murmur went through the briefing hut as the pilots contemplated the meaning of what had just been said. With the ability to escort the bombers all the way to the target, the bombers would protected the entire way. This, Joe thought, will make a tremendous difference. No more easy meat for the *Luftwaffe*. Now they'll have to dogfight over their own territory, greatly increasing the chances for the bombers. It also gave the pilots of the 597th the crack at the *Luftwaffe* that they had been waiting for.

From this point of the briefing, the group-operations officer took over. Joe and the rest of the pilots carefully copied down the information on the backs of their hands and synchronized watches.

When the briefing broke up, Joe walked over to where Rob was in the back of the room, sitting bent forward with his head resting in both hands.

"So, you got the call today as well, I see" Joe said.

Rob looked up through bloodshot eyes and groaned slightly.

"Situation normal... All fu..., oh never mind. Did you get the number of the *Brat* fighter that shot me down?" Rob asked. "I feel like I've been strafed... repeatedly!"

"Believe me, I know just how you feel." Joe rubbed his head and helped Rob up from his chair. "Come on, let's get some coffee."

Checking the roster board, the two pilots located their assigned flight leaders for the mission and made their way out to the flight line

## My Shadow

where *MoJo* and *Shadow* stood fueled and armed by crews who had gotten up even earlier.

"Jeez, you look like hell!" Zeke commented as Joe met him by the plane.

"Yeah, and I feel even worse." Joe replied relieving himself of the vast amount of coffee he had consumed that morning against the sandbags surrounding the hardstand. "But you didn't hear it from me."

Zeke helped Joe into the Mustang and slammed the canopy shut with a bang that shot through Joe's throbbing head like a cannon report.

"Thanks, Zeke." Joe mumbled under his breath.

The fighters took off under radio silence. The four Mustangs of Yellow Flight were directed into position by a flagman on the field. Once again, Joe, flying with Capt. Gauthier and Rob, flying with Capt. Rowan, made up Yellow Flight.

The overcast was thick, though not nearly as dense as it had been on previous missions, and soon all 16 planes were up into a bright sunlight that pierced Joe's eyes like steel needles.

The hard rubber of his oxygen mask chafed against Joe's unshaven face and he increased the flow to 100% hoping to clear his head. Joe would have felt even more sorry for himself if he hadn't known that Rob and many of the other pilots were doubtless in a similar condition.

The Mustangs of the 597th joined up with the rest of the fighters and headed out over the North Sea. The number of P-51s had greatly increased over the last few weeks, but a number of Thunderbolts and Lightnings were still in evidence. The fighters caught up with the bomber formation near the Belgian border and took their escort positions.

Scanning the sky ahead, Joe picked out the B-17s with the triangle 'A' on their massive tails, the Big Friends of the 91st Bomb Group they'd been assigned to escort.

"Well, this is it!" Joe thought as one by one, the shorter range P-38s and P-47s reached *Bingo*, the point where their limited range meant they had to return to base or run out of fuel, "From here on, it's up to us."

Ram-rodding, as veterans called flying escort, could be a long slog, and for the next two hours Joe sat uncomfortably while nothing happened. *MoJo's* small heater was no match for the frigid air at 30,000', and twice Joe had to clear frost from the inside of his canopy, all the while keeping and eye on Yellow Flight and zigzagging along with Gauthier so as not to overtake the bombers.

Alert to anything that moved, Joe watched Rowan and Browning crossing slowly back and forth behind him and it was at this moment that the coffee Joe had consumed that morning began to have its secondary, diuretic effect.

## Robert Brun

Contemplating the situation, the layers of clothing and the almost useless relief tube beneath his seat, Joe's assessment of the machinations such a maneuver would require immediately disappeared when he heard Squadron Leader Davies break radio silence.

"Bandits, bandits, bandits, one o'clock low. 30 plus, punch tanks!"

Adrenaline pumping, Joe's mind suddenly cleared and his headache vanished, now focused solely on the job at hand. His bowels would have to wait.

Joe operated the twin levers that released the auxiliary fuel tanks and loosened up his position on Gauthier's wing, dropping back to cover his leader. Up ahead, the first of the German fighters began their head-on attacks at the front of the bomber stream.

"Yellow two, this is yellow one. Dyer, do you have me? Over." came Gauthier's call.

"Yellow one, this is yellow two, you're clear. Over."

Joe followed Gauthier's perfectly executed split-S and dove into the fray, letting off several short bursts with his machine guns. Pulling hard and rolling to stay in position, Joe's air speed indicator climbed above 400 mph. The 109 Gauthier had been diving on began trailing coolant in a long white line and dropped below the clouds.

Circling around in a climbing turn, Joe followed his flight leader, barely missing the lead bomber and spotted two more Messerschmitts slashing through the bomber box, bringing down another flaming Fortress.

Fifty caliber tracers came from every direction while Joe and his flight leader pursued the two 109s through the formation. Joe and Gauthier closed in on the targets and the two 109s split up, each heading in the opposite direction.

"Dyer, I got right. You get that bastard on the left."

"Left, roger."

Joe pulled back the control column and advanced the throttle to full power breaking left and trying to get on the fleeing 109's tail. The Messerschmitt jinxed and weaved all over the sky, then pulled into a tight turn, banking around to escape his pursuer and get onto *MoJo's* tail in the process. Joe had no intention of allowing either to happen and stuck tight to the twisting, turning 109 who kept pulling into an ever-tightening circle.

After three complete rotations, Joe started to close on the 109 when the German pilot broke away diving sharply for the deck, then gaining speed, pulled into a steep zooming climb, up and over the top of Joe's canopy!

Joe threw the stick over and snapped the Mustang around to follow, banging his head hard against the canopy frame while making a 5g turn that pulled his oxygen mask down across the bridge of his nose.

## My Shadow

Stars flashed before his eyes, but as his vision cleared, the 109 appeared directly in front of him, all set for a zero degree deflection shot, a sitting duck.

Lining up the pipper of his gun sight with the Messerschmitt, Joe fingered the trigger. There were two quick *pops* as four tracer rounds shot out before him and then - nothing!!!

Pulling the trigger several more times while checking the fire control panel to confirm that the gun safety switch was still in the `On' position; Joe saw that it was.

"Con-found-it!" Joe hollered, exasperated, pulling the trigger three more times in rapid succession just to make sure he hadn't overlooked something before breaking off pursuit. No longer able to take part in the battle, the 109 disappeared into a cloudbank.

"The one that got away." Joe thought keying the mic to contact his flight leader.

"Yellow two to yellow one, come in. Over."

"Yellow four here yellow two. What's up Georgia? I thought you had that *Boche* pilot cold, what happened?"

"Guns jammed!" Joe replied, surprised to hear Rob's voice and trying to hide his disappointment. "Not much use here any longer. You want to go after him?" Joe asked.

"Nothin do-in, that's up to you and Gauthier. I gotta keep Rowan's ass outta trouble, that and keep an eye on our Big Friends. Don't worry though, *the Shadow* and the rest of Yellow Flight, will see to it you get home in one piece."

Joe clicked his mic in reply and radioed Gauthier informing him of his situation.

"You hang back there with yellow three and four and cover the bombers. Keep your eyes peeled for anyone trying to sneak in from up-sun. I'll finish up here and form up with the three of you. Out."

"Roger." Joe replied heaving a frustrated sigh.

For the remainder of the mission, Joe could only follow Gauthier and the rest of Yellow Flight, unable to do much of anything, except stew in a broth of his own self-pity. Three more times Gauthier engaged enemy fighters, downing two and increasing his overall score to eleven.

Twice Joe was able to drive off a pursuing 109 from his leader's tail by dashing in and feigning an attack, all the while frustrated that he couldn't fire his guns. Each time the ruse had worked, but the tactic was proving to be a dangerous charade. During the last engagement, Joe flew by so close his wingtip actually clipped off the pursuing German fighter's antenna mast. That had been *too* close!

## Robert Brun

Back at the base, Zeke was up on the wing even before the *MoJo's* propeller stopped spinning and opened the cockpit inquiring about the mission.

"Dammed guns jammed right as I had the Hun bastard all lined up. See what you can do, will ya?" Joe called to his crew chief shouting over the ringing in his ears.

Before he'd said another word, Zeke was out on the wing, produced a screwdriver from his coveralls and removed the access panel to the machine gun and ammunition bay.

"Here's the problem." Zeke called over after examining the guns, "the cartridge belt's jumped the track and the bullet clips parted. Pullin some hard g-turns, were you Lieutenant?"

Joe thought back. "Now that you mention it, yeah. It was pretty hot up there today."

"Damn! That's what I was afraid of. The *SPAM Cans* over at the 357th were having problems like this, I'd talked to one of their crew chiefs and made a few modifications hoping to correct the flaw, but it looks like they didn't hold up under the strain of those high g-turns." Zeke held up a handful of the metal links that connect the bullet belt together. "I'll have to pull the wing apart to see just what happened. Big project, but I should have it corrected before your next mission."

Joe looked over at the frustration on Zeke's face poking around under the wing panel.

"Thanks, Zeke," Joe said, all traces of anger now gone. "Do what you can."

Joe hopped down from the Mustang and watched Zeke and his crew roll the plane over toward the maintenance bay knowing *MoJo* would be well looked after.

After a visit to the field latrine, Joe entered the debriefing hut and gave what information he had to Major Nealson, the S-2 intelligence officer. Nealson's job was to collect information about enemy operations that might prove valuable in future missions. There was the usual short description of flying tactics and details, but today, Joe was particularly brief with his answers. Just as he was finishing up, Gauthier yelled over to where he was standing.

"Dyer! Tomlinson wants to see you in his office, pronto!"

Joe excused himself from the debriefing and headed out the door wondering what else could possibly go wrong today.

Two minutes later Joe entered Tomlinson's office, irritated and not a little bit confused. He had just returned from the days mission hot, sore, hungry, tired and hung over. His combat performance had been less than exemplary and to top it all off, his guns had jammed just when he had that 109 cold, his fifth victory. Now the Colonel wanted to see him.

## My Shadow

Still wearing his flight suit and carrying his helmet and goggles, Joe stood at attention in front of the Colonels desk.

"At ease, Lieutenant," the Colonel said, "and have a seat."

Joe took a chair, nervously fingering his oxygen mask.

"Lieutenant, you've been with the 597th for three months now, and in that time you've exhibited what I consider strong leadership qualities. From your first mission when you lost your flight leader to driving off those 109s and bringing that B-24 back in one piece, and now, today with Captain Gauthier."

Joe sat shifting in his seat, uncomfortable with the praise he'd just received, and surprised that the Colonel knew so much detail about his combat record, but then, he guessed it was the CO's job to know these things.

"In each situation, you've shown that you have the ability to keep your head and think clearly. We need men like that."

"Captain Gauthier has just completed his second tour of duty and has orders to be rotated back stateside as an instructor. As a result, and as of this moment you've been promoted to the rank of Captain and assigned to lead Red Group starting with your next mission."

Joe was dumbfounded. He had had no idea why he'd been called to the Colonel's office, but this was certainly the last thing he'd expected. The Colonel rose and came around to the front of the desk to shake Joe's hand.

"Congratulations Captain Dyer, you've earned it."

Joe stood and shook Tomlinson's hand, still slightly stunned by the event. Then the Colonel spoke again.

"Captain, there's one more thing. You'll be needing a wingman. Give it some thought and let me know if you have a preference by 16:00 today, otherwise you'll be assigned one."

A smile spread over Joe's face, he didn't need the rest of the afternoon, he didn't need another second.

"Browning, Sir. I'd like Lieutenant Browning flying my wing sir."

"Browning..? Lieutenant *Swiss Chee..?"* The Colonel stopped.

Tomlinson sat back down and for a few seconds looked at Joe over the tops of his glasses. Joe had no idea what was running through his CO's mind, but finally the Colonel spoke again.

"Very well then," the Colonel said with a sigh, "Browning it is. He'll start tomorrow. You can inform him yourself. Dismissed!"

Joe gave a sharp salute and left the Colonel's office, grinning from ear to ear.

Back at the equipment hut, Joe entered the crowded room while the rest of the pilots stowed their gear with the usual post mission banter.

## Robert Brun

Excited voices and rapid hand gestures described the day's events as each pilot retold his particular story. Over in the far corner, Joe saw Rob gesticulating wildly, his hands flying in every direction.

"There he is." Rob shouted above the din seeing Joe making his way toward the bench. "Lieutenant Georgia, *Mad-dog* Dyer. You should have seen this guy tear into those *Brat* fighters today." addressing the other pilots there. "I wouldn't be surprised if Göring's put a bounty on his head for what he did out there, breaking up those Jerry S*chwarms* and without even firing a shot! One more and we'll have another *Ace* here at the 597th," and turning to Joe, "You'd have made it too if not for your guns jamming."

Joe ignored the praise.

"Rob, can I talk to you a minute?"

"Sure thing *Mad-dog*, what's your beef?"

"That meeting with Tomlinson, I've been promoted to Captain, and..."

"Hot dog!" Rob barked pounding his friend on the back, "I knew it; I just knew it. Congratulations *Georgia*, Uh, Captain *Georgia*." Rob corrected himself. "No more flying second fiddle, no sir. You'll be a Flight Leader... Who's going to be flying your wing? It'll have to be somebody exceptional, like Taylor, eh no, too twitchy, Harrison, no wait, he's due to go on leave, wait, I know... Cush..."

"Browning." Joe said matter-of-factly.

"Brown...?" Rob stopped in mid sentence and stared at Joe open mouthed. Then after a few seconds he spoke.

"Huh...? *Me...?* Uh... Thanks..." Then lowering his voice, "It would be an honor."

"The honor's all mine, Lieutenant." Joe said putting an arm around Rob's neck, and for the first time since Joe had met him, Browning was speechless.

The next day, Captain Joseph Dyer pulled out onto the airfield as the new Leader of Red Flight. Rob taxied *Shadow* up along side, and with Lieutenants Daniel Taylor and John Cushman making up the remainder of Red Flight. Given the signal, the four Mustangs revved their engines in unison and racing down the field, were soon airborne.

Joe hadn't flown with Rob as his wingman since their first mission together three months earlier, but he liked flying with him again. It was a good feeling seeing the *Shadow* off his wing and he knew he was in good hands.

## My Shadow

The four planes climbed together through the light cloud cover of East Anglia and soon joined up with the rest of the squadron.

**Robert Brun**

# Part II
## Chapter 12
### Mission Scrubbed

*Saturday, February 26, 1944: Great news! Georgia's been promoted and has asked me to be his wingman. What an honor, I think that guy is going places in this man's Air Force - R.B.*

After a week of almost continuous *Ram Rodding*, it was a relief to awaken to see the weather had again closed in and that day's mission was scrubbed. Although the mission of the 24th had been unopposed, tension had been high and the pilots were beginning to feel the strain. Despite the fighter escorts, bomber losses that week had been high. There still weren't enough Mustangs to go around, and the *Luftwaffe* was making the most of that shortcoming.

Eighth Air Force Command had ratcheted up the pressure on the Third Reich by increasing the number of missions and the number of bombers on those missions as well as the number of missions each man had to fly to complete his tour of duty. Originally that number had been twenty-five.

Fly and survive twenty-five missions and you were home free. At the time the odds of successfully reaching that number were only one-in-three, but since January, and with the arrival of the Mustang escorts the number of missions required to complete a tour had been raised to thirty-five. The odds of survival had increased, but so did the number of times you had to face *Lady Luck* and still, she was not always smiling. As expected, the news was not welcomed and it put additional pressure on the already over taxed pilots and crews.

## Robert Brun

Joe lay in bed. It was still dark out and the other men were sound asleep. He envied their unconscious state, but knew there was little chance of his drifting off again so he got up, headed for the latrine, splashed his face with cold water and shook himself awake. Outside the hut, the first rays of dawn were making a futile attempt to penetrate the grey blanket of fog. Joe lit a cigarette while he dressed and exited the hut letting its door bang shut louder than he had intended.

Strolling along the perimeter road, Joe was startled by a jeep that appeared out of the fog carrying a load of tired looking mechanics.

"Morning Cap-n." one Sergeant managed to blurt out above the roar of the motor. "Nice morning for a stroll."

Joe didn't reply, but grunted, shoving his hands deeper into his overcoat pocket and trudged on. Having no specific destination in mind, he found himself at the revetment where *MoJo* was parked.

To cut down on corrosion, Zeke had thrown an old, oil-stained tarp over the cockpit canopy giving the fighter the vague appearance of a hooded bird of prey. The dense, predawn fog had frozen onto the Mustang and a layer of frost covered the metal surface of the wings and fuselage.

Joe savored the peace of the morning, taking another draw on his cigarette that burned his dry throat. The last week of missions had been long and tiring, but the sense of accomplishment it gave seemed to make it worthwhile. He was feeling okay. He liked commanding Red Flight and was pleased his pilots were working so well as a team. Lt. Cushman was proving to be a competent pilot and Lt. Taylor, although a loud mouth, could always be counted on when the time came to get down to work.

Joe ran his hand along the edge of *MoJo*'s wing, his body heat melting the frost into water droplets that trickled down his outstretched arm and he shivered. He had been in theater for two months and already he'd been promoted and made Flight Leader. He was one kill shy of becoming an Ace and the way things were going, he'd soon make that too. Joe had written Mo about his promotion, anxious to share the news, and although he had hoped for a speedy reply, he knew it would be weeks before he heard from her. He wondered what the rest of the folks back home would think of his accomplishments. What about his friends and neighbors and...

...What about his father? What would he have thought were he still alive? Joe hadn't given a thought to the old man since leaving home.

Joe's father, Michael Patrick Dyer, first generation American Dairy Farmer and father of three, what would he have thought of his son? For as long as he could remember Joe had felt he'd been a disappointment to his father, never quite good enough, never quite up to the old man's standards. When he was very young, he had even tried to look to his mother for the approval lacking from his father, but even there, he found

# My Shadow

little support. Maybe it was because of the cautious relationship she had with her husband. His old man could be very intense at times, but would it have been such a great burden to offer a bit of encouragement once in awhile?

When Joe grew older and thought of all the times he had worked alongside his father learning the dairy trade he couldn't recall a single time the man had ever complimented or thanked him. Oh, but he sure remembered the criticism. Nothing he did ever seemed to be good enough or fast enough. Now, ten years later, things had changed. He was a Captain in the United State Army Air Force, a Flight Leader, practically a Fighter Ace. He'd gone and done something good, something worthwhile. Surely his father would have seen that! Only now it was too late, his father was dead and he'd never know any of this. Joe slammed his fist down on *MoJo*'s wing root.

"*Son of a stinkin whore!*" Joe heard the string of expletives coming from inside his plane.

Surprised by the cursing, Joe came around the front of the plane and saw the grey silhouette of a man bent under the tarp and swearing a blue streak. Recognizing the voice and the curses, Joe called out.

"Morning Zeke. What's got you all fired up this early?"

Zeke slowly pulled himself out from under the tarp holding one bleeding hand wrapped in an dirty rag with a look of pure disgust on his face, an expression that instantly reminded Joe of his father and he wish he'd kept his mouth shut. Then he saw a look of recognition on the chief's face and his tone changed.

"Ga-dam manifold pressure gauge has been acting up and I can't seem to get the blasted mounting clips out. Cut my hand on the housing. Hand me that 3/8th box end wrench, will ya Cap?"

Still unsure of himself, Joe turned to the jeep where the chief was pointing and rummaged around in the open toolbox until he found the wrench he hoped Zeke had requested.

With a strange sense of anxiety he hadn't experienced since he was ten, Joe held out the tool meeting Zeke's stern gaze. A frown creased the chief mechanics brow as Zeke screwed up his face then grabbing the tool, broke into his familiar grin.

"Thanks Cap." Spat on the ground, turned and disappeared back under the tarp.

A wave of relief and self-satisfaction washed over Joe as though a tremendous weight had been lifted from his shoulders. As if his eyes were now opened, something inside of him clicked into place and he understood. All this time, he hadn't been wrong and worthless, but it was his father's inability to give praise or accept Joe's desire to please him, make him proud. All this time, Joe had thought he was the failure, that he was never

## Robert Brun

good enough. He'd had it backwards. It had been his old man with the problem.

At that moment Joe felt the anger toward his father vanish, replaced by pity and realization that, up until then, he had hated the man. Hated him for the love, the support, the attention he wouldn't... No, couldn't give. It had been beyond his father's capability. Yet at the same time, Joe now understood what his father had given him, a sense of responsibility and loyalty, the desire to succeed and not quit, all the traits that had gotten him to where he was today. His father had been a sorry man who had missed out on so much of the joy that life and family could bring, but most of all, Joe now understood his life didn't have to be that way. He had moved beyond it. The feeling of uneasiness toward his father faded away and was replaced with a sense of lightness and belonging, perhaps for the first time.

A hard rain began to fall and as Zeke's swearing reached a new crescendo under the tarp, Joe headed across the field toward the mess hut. He was going to enjoy this *day off.*

# My Shadow

## Chapter 13
## Ace Plus

*Saturday, March 4, 1944: Whew, what a week! It seemed like we've been flying non-stop! Finally got a day off to lick our wounds. It's starting to feel like it's either feast or famine. Boredom or terror! - R.B.*

The five days of solid missions that came to be known as "*Big Week*" left Joe and the rest of 597th exhausted. U.S. and British Bomber Command were putting the pressure on German industry day and night. Losses had been heavy, and everyone from Command to Maintenance felt the strain. Things continued to be tense for the rest of February while Bomber Command reevaluated their strategy, so it came as a great relief when some good news finally arrived at the 597th.

At the Officer's Club that evening, Lt. Cushman was in a grand mood. At mail call that morning, Cushman had received the news, through *V-mail*, that he was now a father, his wife having given birth to his son a week earlier. Somehow Cushman had managed to get hold of a boxful of cigars and was running around the Officer's Club passing them out while the other pilots were giving him the usual congratulations and ribbing.

Joe sat finishing up a letter to Mo when Cushman came by with his box of cigars. Happy for the new father, Joe accepted the stogie offered and slipped it into his shirt pocket.

"I'd be careful with that thing," Rob said setting his drink down at the table where Joe sat, "I lit mine up awhile ago... Let's just say a Cuban Cohiba it ain't."

"I'll consider myself warned."

While the two men sat talking, Lt. Harrison came over looking slightly disturbed and cleared his throat.

Lt. Harrison was a tall, thin, easy-going man who had been with the 597th for only a month and had flown three missions so far, each time

## Robert Brun

as Capt. Rowan's wingman. An enthusiastic young man, he nonetheless always looked a little lost. Joe had spoken to Lt. Harrison a few times since his arrival and once after Harrison had received his own personal *Flight of Fancy* from Major Higgins.

"Can I talk to you a minute Captain?" Harrison asked, avoiding eye contact. He looked pale and his hands were trembling when he set down his drink. Rob, noting Harrison's unease, excused himself and headed back over to the bar to join Cushman and the rest of the men.

"Have a seat. What is it, Lieutenant?" Joe asked lowering his voice so as not to be overheard by the other men.

"Captain..." the Lieutenant began, taking a seat and still averting his gaze, "I got a bit of a problem and I hear you'll listen."

Joe was surprised by the statement, but said nothing. Harrison took a deep breath and let it out slowly.

"I've been here for over a month now and flown three missions so far. Tomorrow is number four and... well... I'm... I'm-not-sure-I've-got-what-it-takes... Captain." Harrison blurted out the words in rapid fire as if by slowing, his resolve would have escaped him.

"What do you mean Lieutenant?" Joe asked, probing gently.

"I don't know, Captain, it's kinda hard to describe. When I get up in the morning, before a mission, I wake in kind of a mild panic. Once up, everything starts out okay, but then when I head into the mission briefing, my stomach starts to flip-flop. I keep holding myself together as best as I can on the way to the flight line... Then, well, let's just say I have to hit the latrine... fast! Yesterday I almost soiled my flight suit." Harrison's face turned bright red. There was a long pause and gulping down the rest of his drink he continued.

"Then there're the missions themselves. The routine of climbing up into formation keeps me busy, and I like the flying and all, but when I saw those 109s appeared on the horizon the other day... I was breathing so hard I thought my heart was going to pound out of my chest and sweat was pouring off my forehead. Dammit Captain," Harrison banged his fist on the table drawing attention to himself then lowering his voice. "I just wanted to get out of there! Turn my plane around and run, head back to base, make up some mechanical excuse to abort and take the consequences." Harrison lowered his head.

Joe sat quietly for a moment, processing what he had just heard. He took a sip of his beer and set it down on the table between them.

"But you didn't, did you?" he said calmly.

"Sir?" Harrison said, looking up puzzled.

"You didn't abort, did you?"

Harrison was silent.

"Lieutenant, take a look around this room."

## My Shadow

All around the room men were talking, drinking and chiding Cushman about fatherhood. Someone had sketched up a drawing of a girl in a compromised position and pinned it to the dartboard where the men were placing bets and throwing darts in celebration of Cushman's manhood. It all looked very festive to Harrison and only served to make the Lieutenant feel that much worse.

"Lieutenant, there's not a man in here who hasn't gone through exactly the same thing you're going through right now," and then after a pause, "myself included... And if he says he didn't, he's a damned liar!"

Harrison looked up surprised, glancing around the room again with this new insight. He watched the veteran pilots the ones he looked up too laugh and go on drinking and smoking Cushman's bad cigars. They all seemed totally at ease, but, as if blinders had been removed, Harrison noticed, for the first time, something he hadn't seen before. Something just below the surface of the swagger and self-bravado... A strange unease, a tick, a twitch or a quirky gesture, and the eyes, always moving, it was subtle, but evident, if you knew what to look for. Buried deep, but it was there just the same.

"I guess I never thought of it that way before, Captain." Harrison said, his eyes bloodshot.

"We may fly our missions in separate planes, but we are all one team here. No one has to go through this alone... unless he chooses too." Joe glanced over at Rob then sat back quietly hoping his words had gotten through to the Lieutenant.

"Now, get some sleep, Lieutenant." Joe said leaning across the table. "We've got a mission in the morning."

"Yes-Sir. Thank you Sir." and he got up to leave.

Joe watched the young Lieutenant head to the bar, where he congratulated Cushman then left the Officer's Club.

After Harrison left, Joe pulled Mo's picture out of his pocket. The photo was getting creased and dog-eared and the top center was worn away from repeatedly slipping it under the gun-sight.

"Where the hell did THAT come from?" Joe asked the photo, thinking back on the conversation he'd just had with Lt. Harrison.

"Four months ago I was that same snot-nosed *newbie*, wet behind the ears, and now I'm the Flight Leader giving fatherly advice."

Joe shook his head and slipped the photo back into his pocket wishing he'd hear something back from her.

Finishing his drink Joe walked over to the bar where the rest of the pilots were still buying Cushman drinks and smoking his cigars. Cushman was starting to look a little glassy eyed when Joe approached.

*"Cap-pin..."* Cushman said, with a pronounced slur.

## Robert Brun

"I been flyin combat for three munts now and I ain't got one kill to my credit. Tomorrow, we got an-nudder mish-on and I'm gonna get me a Jerry plane this time for sure. One for my boy.., kind of a birt-day present."

"Yeah, Lieutenant. That'll be real nice, now why don't you go and get some sleep?"

"Sleep, dat's a reeeally good idea." and the words no sooner left his month when Cushman slumped over onto the bar.

Catching him under the arms, Joe and Rob lifted the Lieutenant to his feet and carried him toward the door.

"Let's get Papa to bed."

The next morning arrived much too quickly and after the usual morning routine to which, Joe had now become accustomed, he once again found himself at another mission briefing. Tomlinson entered the hut in his usual form.

"All right men, listen up." Tomlinson barked across the room taking the platform at the far end of the hut.

"We've got a very special mission today."

The curtain covering the mission map was withdrawn and this time the red yarn stretched all the way to Berlin.

"Bomber Command has ordered that it's time we take the battle right to Hitler's doorstep and hit him where it'll hurt the most, the Nazi capital: Berlin!"

The mumbled voices that sounded throughout the briefing room were sudden and intense. Each pilot had known that this was coming, and now emotions were raw, a mixture of excitement and fear.

"Berlin? The big 'B." Joe said the word out loud and stared up at the map before him.

"Okay...okay, quiet it down, men." said Colonel Tomlinson, again addressing the pilots assembled.

"Shut yer yaps!" shouted Group Leader Davies, and the room suddenly went quiet.

"Although the Brits have been hitting the Nazi capital for awhile now, this is the first time for the 8th. Jerry's pulled back most of their frontline fighters so they'll be waiting. I'm not going to pretend this is going to be a milk run."

Joe still couldn't believe it, Berlin, and in broad daylight. Not another word was uttered from the pilots assembled throughout the rest of the briefing. The Colonel's aide then took over filling in the details of the mission and the weather officer finished up giving his report on conditions over the target: 8/10th cloud cover, but believed to be breaking up.

## My Shadow

Later that morning, Joe sat in the cockpit of the Mustang waiting for the signal from the tower. Today's mission was to be their deepest penetration into Germany yet, but the two 108-gallon auxiliary fuel tanks hanging from *MoJo's* wings assured that the bombers would have company both ways.

Despite the constant bombing of the aircraft factories throughout February, recent reports indicated that the number of enemy fighters had increased, and Joe knew the Germans would aggressively defend their capital. They were in for a rough trip.

Taxiing out onto the perimeter track, Joe passed Lt. Harrison climbing into his fighter and gave him a relaxed salute. Harrison returned the salute followed with a big grin. Further along the flight line, Lt. Cushman was having a harder time of it. He seemed to be experiencing some trouble getting settled into his plane, but managed a shaky salute when he noted his Flight Leader's gaze.

Rob taxied *Shadow* up alongside Joe at the end of the field and signaled he was ready for takeoff.

All takeoff instructions continued to be given by what had now become a sophisticated set of hand signals. No sense letting the Jerries know when they were coming, they'd figure that out soon enough.

Joe looked over toward the control tower and watched the red flare, indicating a "GO" for today's mission, fired into the air.

Catching up with the bombers had been easy, the Mustangs of the 597th picking up the white contrails long before seeing the bombers themselves. It seemed with each mission, the number of planes was increasing, and today the stream stretched all the way to the horizon.

Reaching the Dutch coast, Red Flight held together nicely, taking up their position above their assigned bombers. The tail markings *square "B"* and *square "D"* indicating the 95th and 100th Bomb Groups were the 597th's Big Friends for today and Red Flight began flying their lazy S-turns along with the rest of the fighters, hundreds of them. They'd certainly come prepared for their first visit to the big "B."

Unable to fly as high with a full bomb load, the faster Liberators could be seen below the Fortresses crossing into Holland. As was usually the case, the *Libs* soon turned off on their own, heading for a separate target and taking their share of Little Friends with them.

During a mid-course correction to avoid the heavier flak around the city of Bremen, random bursts of light flak began appearing below.

"Here we go gentleman. Wingmen, stick close to your Flight Leaders and keep your eyes open for the fighters."

*Flight Leader,* Joe still liked the sound of that. It had been a month since his promotion, and Red Flight was proving to be an efficient fighting unit. His three pilots worked well together almost anticipating each other's intentions. Joe was proud of his men and once again, thought of his father. It was also the news about Cushman's son that gave Joe a feeling he could start over with his own father even though he'd been gone for years. A chance to set things straight in his own mind, to relive his childhood vicariously and rebuild the relationship he never had with his own father, maybe one day, even with his own kids.

The cloud cover up ahead continued to thicken when the bombers approached the German border and ten minutes later, nearing Berlin, the first enemy fighters appeared as groups of tiny black specks climbing up from below.

Joe advanced the throttle and pushed the nose of his Mustang down to intercept the approaching fighters. Small flashes appeared along the cowling of the 190s, the German pilots using their 13mm machine guns to acquire their targets before opening up with the four, more powerful 20 mm wing mounted cannons.

"Release tanks. Keep together and let's break-em up." Joe radioed Red Flight while closing in on the approaching Focke-Wulfs.

Opening fire at one thousand yards, the flight of four 190s split into two groups, diving left and right below the lead bombers.

"Mad-dog Red three and four, take left. Rob, you stick with me."

Red Flight split into two sections in pursuit of the diving fighters. Like clockwork, Joe and his wingman followed the other two 190s diving away from the bomber stream. Glancing in his mirror, Joe made note of Rob's presence and rolled left after the closer of the two Focke-Wulfs.

Lining up for a better firing angle, Joe cut sharply inside his wingman's path and pulled up right in front of *Shadow* forcing Rob to chop powder to avoid a collision. Firing off a quick burst, the rear 190 began trailing smoke and fell into a cloudbank.

"Just a probable." Joe thought knowing he wouldn't get full credit for the kill.

Joe leveled his plane and Rob pulled in right beside his flight leader's wing.

"What was all that about?" Rob inquired of Joe's radical maneuvering. "You nearly got a trim."

"Man at work." Joe remarked, shrugging off the comment and continued to scan the sky for fighters.

Up ahead, Joe could see the remainder of Red Flight circling out ahead of the bomber stream. They were nearing the I.P. and Lieutenants Taylor and Cushman had leveled off beside the lead bomber when Joe spotted a solitary Bf-110 sitting at about 1,200 yards, outside the range of the Fortresses' gunners.

## My Shadow

The Bf-110 *Zerstörer* (Destroyer) was designed in 1934 as a long-range fighter intended to escort German bombers on missions too long for the short-range 109s. The plane had preformed well at first, dominating the Polish and French Air Forces, but once matched against the British Hurricanes and Spitfires, these escort fighters proved to need fighter escort of their own. Now, having been reassigned as night fighters, it was unusual to see one in broad daylight.

Joe's curiosity about the presence of this obsolete aircraft was soon quelled when twin streaks of smoke appeared from beneath its wings.

"Rockets!" Joe said aloud, the two trails of smoke now heading toward the bombers with Joe and Rob directly in their path.

"Red two, BREAK!" Joe shouted into the radio and without hesitation the two Mustang split up, the streaking projectile passing right through where they had been just seconds before. Joe recovered quickly and turned his head in time to see one of the rockets slam right into Cushman's Mustang and explode.

Taking the full impact of the hit, the Mustang broke apart, its right wing separating from the fuselage and the flaming remains spiraling down out of control. It was gunnery practice and Ben Howard in his burning Warhawk all over again, but this time Joe could hear Cushman's screams in his earphones, then nothing but static and the drone of his own Merlin. Joe stared in stunned disbelief.

After only a moment's hesitation Joe depressed his radio transmit button speaking slowly and deliberately.

"Browning, Taylor, stay with the bombers." and not waiting for a reply, Joe banked *MoJo* around and took off in the direction of the lone enemy aircraft.

The twin-tailed Messerschmitt was no match for the speed and agility of Joe's Mustang and soon he had caught up with the 110, getting it right where he wanted it.

Closing to within 100 yards behind the much slower German fighter, Joe placed the pipper on the center of the port engine nacelle and fired one, then another and finally a third short burst into the fleeing *Zerstörer*.

Tracer, armor-piercing and incendiary rounds tore into the twin-engine fighter and pieces of the wing began to fly off, Joe's bullets hitting their target. The left engine began smoking and flames had engulfed the entire wing when the canopy flew off, and a single figure emerged tumbling from the crippled aircraft.

Banking his fighter to follow the burning 110, Joe saw the yellow-brown parachute stream out from the falling figure and open with a snap. The German pilot now hung helplessly before him.

## Robert Brun

Observing the gentle rocking motion of the swinging pilot, a perverse image of a baby's cradle appeared in Joe's mind, the cradle of the son Cushman would never see, a child that would never know his father. Joe's rage grew out of control and he closed in on the dangling pilot. Like a scratchy old phonograph, a lullaby echoed through his head.

"*Rock-a-by ba-by, on the tree top!*"
Joe lined up the pipper on the hanging figure.
"*...When the wind blows, the cradle will rock...*"
Joe tightened his grip on the control stick trigger.
"*...When the bough breaks, the cradle will fall...!*"

...Joe's mind screamed! The realization of what he was about to do wrenched him back to sanity and instantly he threw the control column over missing the helpless pilot by only a few feet.

Looking into the mirror, he watched his prop-wash collapse the parachute canopy and could see the terrified look on the pilot's face, the speed of his fall increasing. Then the parachute caught once again, reopened and the German pilot drifted down out of view.

The Bf-110 had been his fifth kill. Now officially an *Ace*, Joe felt only disgust.

Collecting his thoughts, he eased the stick and lowered his oxygen mask to wiped the sweat off his face with a shaky hand.

"Georgia... Captain... HEY, *MAD-DOG!* We've still got company... Over"

Joe's stupor was interrupted by the sound of Rob's voice over the radio headset.

"Roger, Red two." Joe shook the scene from his mind and rejoined the rest of the fighter escort not yet fully recovered from the event of the last few minutes.

Back among the stream, fighter planes dove and turned from every direction. Bombers were dropping out of formation and with no shortage of targets for either side, brown and white chutes were everywhere.

Flying straight into the battle, Joe ignored the tracer fire and selecting another 190 with a 30-degree deflection shot, raked the fighter from end to end. Number six.

Not missing a beat, he singled out a second Focke-Wulf and just as quickly dispatched that one as well.

Then, just as suddenly as it had started, it was over. The enemy fighters disappeared. The bombers had reached the I.P. and lined up for the bombing run to the target.

## My Shadow

The flak over the German Capital was the worse he'd ever seen and Joe counted five more Fortresses hit over the city. He wanted to do something, anything, but against flak there was nothing to be done. Still the bombers flew on, straight and level, perfect targets for the German 88 gunners below.

The fighters picked up what was left of the bomber force once they had circled after dropping their loads. Relieved of the weight of the bombs, the Fortresses broke away from the city and reformed into defensive boxes, filling in the slots vacated by those planes that had been shot down. Many showed battle damage, some trailed smoke, while others had propellers feathered on shot out engines. It had been a hard fight, and it was far from over.

With a cold detachment, Joe went about his business and by the time he had reached the Dutch Coast, there were two less Focke-Wulfs to worry about.

Without a word to Zeke, Joe climbed out of his plane and headed toward the debriefing hut. Rob caught up and silently followed his Flight Leader into the hut.

The S-2 officer, Major Nealson, was seated behind the desk staring up at Joe when he entered the room. Joe returned his stare then Nealson spoke up.

"Did you get the recall order Captain?"

All the color drained from Joe's face and for an instant Rob thought Joe might pass out.

"Recall order?" Joe asked in a barely audible tone, "What recall order?" turning to look at Rob who was standing beside him. Rob shook his head.

"The mission was recalled just after crossing the Dutch coast. Weather over the target had deteriorated, apparently you and a wing of the 3rd air division all missed the signal."

"Sir, I..." Joe stammered.

Throughout the rest of the interrogation, Major Nealson asked the standard debriefing questions. Joe's answers were short, curt and, at times, almost rude to the point that even Major Nealson took note of Joe's uncharacteristic behavior. Finishing his debriefing, Joe saluted and swiftly exited the room.

"Lost a man today." Rob explained to the S-2 officer as he met the man's puzzled gaze.

With the unstable weather during the winter, mission recalls were fairly common and particularly hard on the men. Not only was the anxiety level the same as a full mission, a recall was not credited towards a pilot's

## Robert Brun

mission total so nothing was gained, but Joe had missed the recall! How had it happened? And now, one of his men was dead. Joe was conspicuously absent at dinner that evening and breakfast the next day.

By Monday, with Cushman's death still weighing heavily on his mind, Joe found himself, once again sitting in a mission briefing awaiting the Colonel's arrival. He had kept to himself over the weekend choosing to remain in his quarters and was still in no mood for company when Rob entered and took the seat next to his. The two men exchanged glances, but remained silent, then the Colonel arrived.

"Two days ago, Saturday, March 4th, two squadrons of the 95th Bomb Group and one from the 100th with fighter escort dropped the first U.S. bombs on the Nazi capital of Berlin. Today, we're going back again... in force."

"Just like it was two days ago." Joe thought. A sense of *deja vu* enveloped him and continued, unabated throughout the rest of the briefing, replaying the events of the last mission over in his head for the thousandth time. The rockets, the explosion, the screams, had he made the right decisions? How had he missed the recall?

Joe was awakened from his rumination when he heard the adjutant officer bark:

" That's all men, Dis-Missed!"

Getting up from his seat and still in a fog, Joe started toward the door when the Colonel's aide approached.

"Captain Dyer... Colonel Tomlinson wants to see you in his office right away."

Joe glanced over at Rob who said nothing.

Moments later, standing before the Colonel's desk, Joe was instructed to take a seat.

"Captain," the Colonel began. "I understand you lost a man on the last mission."

"Cushman sir, that's correct."

"Captain, it's never easy to loose a man, especially one directly under your command, but it happens, and it will happen again. How are you handling it?"

"I'm *fine* Colonel."

Tomlinson took a long look at Joe as though sizing him up. Then he spoke again.

"It's hard Captain. I know, and it never gets any easier. Be thankful it doesn't. You learn to live with it though God only knows how."

Joe didn't say a word.

"As a Flight Leader, it's your job to make tough decisions, not only for yourself, but for the men under your command and sometimes

## My Shadow

that's hard to accept." There was another pause. "Captain Dyer, you're one of my best pilots and as Flight Leaders, I can't afford to loose you." The Colonel sat looking at Joe for a long moment, then continued.

"Captain, I've assigned Captain Rowan to lead Red Flight on today's mission."

"Sir?" Joe asked incredulously. "Begging the Colonel's pardon, but I don't see the need..." Joe started to protest, but Tomlinson cut him off.

"Captain, you've had a couple of rough weeks. What you need now is a break."

"Colonel, I feel that I can better serve my men, the Group and the Squadron by continuing to fly."

"Captain, I know what you're going through and..."

"DO YOU?" Joe's voice grew to a shout before he realized to whom he was speaking.

YES Captain, I DO!" Tomlinson shot back narrowing his eyes. Then he lowered his voice looking directly at Joe.

"...Every time one of those pilots doesn't come back... and there have been plenty who haven't." An icy silence hung in the air.

"Joe, for far too long now you've been pushing yourself hard..." The Colonel paused and took a deep breath then continued.

"...Captain, I'm grounding you until further notice. You'll be assigned to squadron administrative duties in the meantime."

Joe stood defiantly. "Will that be all Colonel?"

"No! One more thing captain, I want you to see Doc Lewis. I've instructed him to give you something to help you sleep. Now get some rest. That's an order. Dismissed"

"Sir I..." then Joe thought better of it. "Yes Sir."

Of the over 700 bombers that had flown to Berlin on March 4 and 6, 1944, seventy-four were lost along with eleven escort fighters. 751 men gone, Lt. John G. Cushman had just been one of many, but he was the only one Joe knew by name.

# Robert Brun

## Chapter 14
## Transitions

*Sunday, March 26, 1944: We lost Cushman on the last mission, a freak shot from a Brat rocket. Georgia's taking it real hard. I'm sure gonna miss that big galloot – R.B.*

Joe didn't know what Doc Lewis had given him, but it did the trick, sleeping through the night deep and dreamless. He awoke the next morning to the sound of the squadron taking off without him feeling both dejected and alone. Without flight duty, he lay there not knowing what to do with his time. He couldn't get back to sleep, wasn't hungry and his new duties didn't start until tomorrow. He couldn't even bring himself to write Mo! What was he going to tell her; he'd screwed up and gotten a man killed? Joe rolled over and stared at the wall. For another hour, he lay there feeling sorry for himself, replaying the events of the last mission in his head.

Making Ace, leading men, overconfidence and pride before his fall all had played a part, but what had he thought, that he could somehow control the chaos around him? Cushman was dead and yes, he was to blame, but by following that logic, so was North American Aviation for building the Mustang, *Bayerische Flugzeugwerke* for building the 110, the maintenance crews for maintaining the plane that put them there, the bullets, the rockets, Rob for dodging the rockets, the cooks for preparing his meals, and the Doc for keeping him healthy enough to fly, and so the logic breaks down. They were all to blame and at the same time blameless. Responsibility within chaos, was that the true job of the warrior?

What bothered Joe the most was the effect this was having on him. Hatred and revenge had been feelings foreign to him, but now they were real. For the first time, he had wanted to kill, premeditated and with malice aforethought and it scared the hell out of him. What had he become? He lay there listening to the sounds of the base activity until he'd had enough, then dragged himself out of bed, washed and dressed.

## My Shadow

Entering the mess hall relieved to see most of the pilots were gone, Joe sat away from his usual table. The men there seemed to respect his decision by keeping their distance. While finishing his coffee, Major McNulty the base Chaplin, a quiet man in his early forties with a good heart and open mind, came over and sat down.

"Captain, I understand you lost a man on the Berlin raid?"

Joe stared into his coffee cup without replying.

"Captain, I know some of the pilots here don't have much regard for anyone who doesn't fly combat, but we men of the Cloth weren't always so. You may find I can understand a good deal more than you might imagine."

Joe looked up confused as the Chaplin slid up his sleeve exposing what was clearly a healed bullet wound of a large caliber.

"I've known loss myself." He added. "And if you want to talk about anything, you know where to find me." The Chaplin rolled down his sleeve, got up and left. Joe didn't move, but sat thinking. He knew death was part of war, so why had Cushman's hit him so hard?

While he sat, the mess crew cleared the morning meal and immediately began preparing for the next. Noticing the activity, Joe looked up. Having never seen this before, it hadn't occurred to him what was involved in the daily feeding the 300 plus men stationed at Abington. The pilots flew the planes, the mechanics kept them flying and the staff officers ran the works, but everyone had to eat. Impressed not only by the speed, but also the efficiency of the crew, Joe began to see how it all worked. Flight Leader or not, they were all just a cogs in a wheel.

Although not a particularly religious man, Joe thought perhaps the Padre was right, just speaking with someone who would listen without judgment might help him sort out the guilt he was feeling over Cushman's death. He met with the Chaplin twice that week and as the days wore on and his resentment at being grounded eased, so did his guilt. Joe began to relax and accept his situation. The Colonel had been right; the strain of continuous combat had gotten to him more than he had wanted to admit.

Because of the shortage of planes, *MoJo* had been assigned to Capt. Rowan while Rowan's plane was undergoing an engine replacement. Capt. Rowan was an experienced pilot and he knew the *MoJo* was in good hands, but couldn't help feeling uncomfortable about it. Even though Zeke would disagree, he felt like it was *his* plane and didn't like the idea of someone else at the controls. Just the thought reminded him of *Doug*.

During the remainder of his ground time, Joe was far from idle. He took on his new assignment, overseeing the mission assignments and various administrative duties with the same dedication he had as a Flight

## Robert Brun

Leader. There was still the strong discomfort at not being with the other pilots, but he was discovering the other side of this war and found by far the hardest part was the waiting. Watching the squadron take off without him was bad enough, but far worse was counting the planes upon their return... and coming up short, particularly until *Shadow* touched down.

That week, two planes had failed to return. The first had diverted to the bomber base at Halesworth with engine trouble and the other, returning with battle damage, had ditched in the channel (the pilot of the second being picked up by a fishing trawler a few hours later). Until he had learned their fates, Joe had remained on edge. He was developing a new respect and appreciation for the difficulty of Colonel Tomlinson's job and understanding of his own responsibilities as Flight Leader.

After Friday's mission, while Joe stood on the control tower platform watching the planes returning through binoculars, he spotted *MoJo* coming in uncharacteristically high and fast. Bouncing several times on touchdown, the plane nearly ground-looped before skidding to a stop at the end of the runway. Running from the control tower, Joe followed the *Meat Wagon* over to the taxiway and watched the orderlies extract Capt. Rowan from his plane, blood covering his arm and left side.

"Took a slug in the upper arm." Doc Lewis said while Joe looked on. "Looks like it might have shattered the bone. He'll be out of action for quite a while."

"*Yeah, and it hurts like a son-of-a-bitch.*" Rowan hissed through gritted teeth. Carried past on the stretcher and lifted into the waiting ambulance, Rowan was given a shot of morphine to ease the pain.

After the trucks had pulled away, Joe walked over to his plane and climbing up on the wing, ran his hand along her olive-drab colored aluminum skin. There, just beneath the side canopy, was a small, perfectly round hole no bigger then a pencil. Peering into the cockpit, Joe saw the results of this innocent looking puncture. Capt. Rowan's blood covered the left harness strap and had pooled in the seat pan below. The red, opaque liquid's sickly sweet odor filled *MoJo's* cockpit.

"That's all it takes," he thought, "one lucky shot." Joe closed the canopy and climbed down off the wing.

While heading toward the base operations building later that afternoon, a mangy looking canine came racing out of the mess hall with half a roasted chicken in its mouth followed closely by Cookie, one of the mess crew, wielding a large chef's knife.

"Captain! Grab that four legged son-of-a-bitch will ya?"

Sticking out a leg, Joe blocked the animal's escape and swiftly grabbed the cur by the collar.

## My Shadow

Cookie came running up and relieved Joe of the whimpering pooch, carrying the dog, chicken and all, off by the scruff of the neck back in the direction of the mess building.

"Howdy, Georgia." a familiar voice called from behind.

"Not flying today?" Joe replied, surprised to see Rob, then followed with, "He's not going to hurt it is he?" watching the angry cook carry the whimpering dog back inside the mess hall.

"Are you kidding? That hound gets away with murder around here. Those two just like to go at it every so often. Besides, it's his dog."

Joe had to laugh at the irony.

"As for your first question: No, I thought I'd give some of the other guys a chance at the glory today. Besides, this way I get to keep an eye on you tomorrow."

Joe's expression brightened, "You mean..?"

"I don't know nothin," Rob said putting up both hands, "cept your name's on the board for tomorrow's mission. Oh, and Tomlinson wants to see you in his office." Rob winked with a grin.

The Colonel was shuffling through papers on his desk when Joe entered and stood at an uneasy attention.

"At ease, Captain."

Joe relaxed, but only a bit.

"I've spoken to Doc Lewis and he's cleared you to return to flight duty. You'll resume your assignment as Red Flight's leader and starting next week you'll be relieving Major Davies as Squadron Commander."

The Colonel's words surprised Joe. This meant he would now be commanding the entire squadron, responsible for four flights, sixteen planes, and sixteen pilots. Joe flushed at the news.

"Lieutenant Harrison has been assigned to you as your number four, Lieutenant Cushman's replacement."

The mention of Cushman's name caused Joe a twinge of guilt that he hoped didn't show.

"Briefing is at 06:30 tomorrow sharp! Captain, It's good to have you back."

"It's good to be back Sir." Joe stood, saluted and turned to leave, then stopped. "How's Captain Rowan?" he asked.

"Doc's removed the slug and set the arm. He was messed up pretty bad and still in a lot of pain, but he'll pull through. I'm afraid he won't be flying again anytime soon."

Joe paused to consider what was just said and wondered how he'd feel if not able to fly. His earthbound duties had confirmed what he'd suspected; in the air was where he belonged. He closed the Colonel's door.

## Robert Brun

The next day, Joe was back behind the controls of *MoJo*. The inside of the cockpit had a faint antiseptic odor and there wasn't a trace of blood anywhere. The hole in the side panel was gone and a fleshly painted plate of new aluminum had been riveted in its place.

"Zeke, you are truly amazing." Joe thought.

For the next two weeks, Joe led the squadron through multiple missions and each flight rebuilt his confidence. The P-51 Mustang had by now replaced most of the other squadrons' fighters and long escort missions to the Third Reich's Capital were becoming commonplace. The weather, however, continued to be uncooperative.

The winter of 1943-44 had been one of the worst on record and the spring was turning out to be just as bad. As often as not, missions had to be scrubbed. By the end of March, when the weather finally began to clear, Lt. Harrison, had settled in and was filling Red Four position nicely.

Joe still had not heard from Mo despite his writing to her now that he was flying again. What was going on? He wanted to know. With each mail call came an expectation of his own *"Dear John"* letter. Would that be preferable to the waiting? The anxiety of being so far away was getting harder to ignore.

During stand-downs, when the weather was so bad that even the birds were walking, Joe found it difficult to occupy his time. Rob, of course, had no trouble, his nose always buried in his sketchbook or working on one of his numerous paintings that, by this time, adorned the walls of almost every building throughout the base including the maintenance hangers and even some of the latrines. Somehow he'd managed to scrounge together enough paint, brushes and a makeshift easel, and would often be seen out painting on even the foggiest of days, said he liked the *atmosphere* it created. Although Joe still hated fog, the *F-word* as he called it, he admired Rob's atmospheric paintings just the same and had to admit, the guy was talented.

Eventually, after weeks of searching, Joe and the 597th finally did find their distraction out on the base's airfield playing, of all things, soccer or *"football"* as the Brits called it. Because it required no equipment, it could be played in even the worst weather, which naturally accompanied any spare time they now had. Joe and a number of the other pilots and ground crews would assemble in mud, fog or rain kicking a tattered old leather ball someone had stolen from the nearby town. During these games, military rank no longer mattered all players were equals.

Using 55-gallon fuel drums stacked with two-by-four crossbars as goals, the games went on for hours with little regard for the score or

## My Shadow

physical well being. Cuts, scrapes, and bruises were the norm and Major Lewis, the base doctor, was kept plenty busy.

Once the Colonel got over the fact that they weren't playing *American Football*, Tomlinson saw these minor injuries as a fair exchange for the morale boost it gave his men. After a game, they would return filthy, exhausted and thoroughly relaxed. On days when they played - and with the early spring weather of '44, it was often - Joe slept soundly and awoke rested.

March winds gave way to April showers as the weather in East Anglia went from cold and snowy to cold and wet, but it was always the same... *cold*. Frozen ground thawed and mud was everywhere and in everything.

From the moment you awoke in the morning to the minute you went to bed, the sticky brown stuff accompanied you, following like a slug's slimy trail. It was on your boots, it was in your bunk, on your pants, in your cockpit and at times, when the drains got plugged, even in the showers. Some felt it was in the coffee too.

Joe hated the mud even more than the fog, though perhaps at the same time, welcomed it as a focal point for his growing anger and frustration. It was getting on to nine weeks and there was still no word from Mo. The waiting was getting to him and he knew it.

By mid April the weather cleared and Joe no longer had time to think about much more than the missions. Having fully recovered from the devastation of "Big Week," the 8th Air Force had shifted, the focus of the bombing offensive and escort missions to Berlin were becoming routine. With Mustangs in sufficient numbers to escort the bombers properly, Joe and the rest of the 597th had little rest, flying three out of every four days for almost a month.

In addition to the raids on Berlin, escort missions over France to hit the marshaling yards, disrupting German transportation and creating general mayhem continued. It was clear that Bomber Command was softening up the Continent for something, but just what, no one knew for sure. One thing was certain however, fewer German fighters were appearing to harass the bombers and by early May it was clear the allies were gaining control of the skies over France.

Since his promotion to Flight Leader and now to Squadron Commander, Joe had learned first hand the hard lesson; with rank comes responsibility. It had been a short, but painful education, and the cost had been high. With Cushman's death, Red Flight had lost a fine man, a new father, and a friend, but it also forced Joe to realize that it wasn't just *his* men, *his* friends who were dying, but everyone's and on both sides.

# My Shadow

## Chapter 15
## London Leave

*Thursday, June 1, 1944: Finally a break in the routine. Off to London for R&R. Can't wait to see the sights and do some sketching – R.B.*

The three-day pass was a godsend after the weeks the 597th had been through and Joe was looking forward to this break in the routine. Just the change of scenery alone would do him good so for the next three days, Joe was determined to forget the last few months. Despite not hearing anything from Mo, Joe decided, while on leave at least, he was going to put everything out of his mind and just relax. He'd earned this break and planned to take full advantage of it.

Having caught the first lorry to the train station, Joe and Rob smoked and waited impatiently with the rest of the GIs on leave.

"I haven't been to London since we got here." Rob said with excitement as he peered down the rails in anticipation of the train's arrival. "And I didn't get a chance to see anything before we left for this place."

"I know what you mean, but the first thing I'll be doing once I get there is to soak in a hot bath. Then I'm gonna buy myself a decent meal at the best restaurant I can afford. No SPAM for this soldier for awhile!"

Moments later, the smoking locomotive of the London train screeched into the station, and Joe and Rob boarded the train.

The trip into London was much like the one down from Scotland had been, the train's cars packed with soldiers. This time, however, the mood was very different. Listening in on some of the chatter around him, Joe noticed that these were not the conversations of young, eager boys wondering about their immediate future, but rather the musings of men trying to forget themselves and their duties for a while.

## Robert Brun

Joe and Rob took up much the same position as they had some months ago. Reclining in the seat with his feet resting on his kit bag, Joe was surprised by the sudden fatigue that washed over him and without a further thought he pulled his cap down over his eyes and promptly fell into a deep sleep.

"*London...Next stop, London!*" came the conductors voice. Joe opened one eye and took in the scene outside the coach window. It was mid-afternoon and despite the ubiquitous overcast, the station was buzzing with people.

Nudging Rob with the toe of his boot, Joe rose from his seat, tossing the small kit over his shoulder.

"Looks like we made it buddy. This is our stop."

Rob groaned and rubbed his eyes. "You have NO idea the dream you just woke me out of, you bastard, but I'll tell you one thing, you owe *Miss Grable* an apology."

Joe chuckled and the two pilots exited the train pushing their way through the busy station where soldiers of every size, description and nationality bustled shouting in as many languages.

The dialects of Frenchmen, Poles, Australians, ruddy Scots, dark skinned Africans, and several others Joe couldn't identify greeted his ears, each expressing a feeling of friendship and camaraderie. This war was truly a worldwide conflict.

A large contingent of *Free French* joyously greeted the two pilots with rib crushing hugs and a kiss on each cheek as they passed. A small group of Poles recognized them as fighter pilots and gestured a pantomime of shooting down an aircraft to peals of laughter and sounds of *'Ratt-tat-tat-tat!'* It felt good to be surrounded by such merriment for a change.

"Where're you going first?" Rob inquired as the two stepped out into the busy London street.

"I'm going to check into the hotel and get me that bath." Joe proclaimed.

"Well, not me!" Rob replied. "I'm gonna see as much of this city and do as many things as I can over the next 72... er" and, looking at his watch, "68 hours. For the next three days, I'm strictly a tourist. It's pictures and souvenirs for me. So if we don't run across each other, I'll see you back at the base in three days." And with that, Rob whistled up a passing taxi, and took off in a cloud of exhaust.

Joe, feeling somewhat abandoned for the moment, soon fell under the spell of the strange city that lay before him. Having grown up in rural Georgia, he still felt like a fish out of water so pulling a slip of paper Lt. Taylor had given him out of his pocket he read the name: *'Regent's*

# My Shadow

*Palace, Piccadilly.'* With a sigh of uncertainty, Joe raised his hand as he had seen Rob do and whistled for a cab.

To a chorus of auto horns and screeching tires, a beat up black London taxi executed a rapid U-turn and promptly pulled up alongside the curb, the driver hopping out and opening the passenger door.

*"Where to, mate?"* the cabbie asked

"Regent's Palace, Piccadilly" Joe replied.

*"Go-wall!"* the cabbie replied. *"Good luck wit dat. Been drivin Yanks dare oll day now, I 'ave. Place be right full up by now I 'spect, but it's your shillin."*

From the rear seat, Joe said nothing while he looked out the window at the sights of this strange city until the cab pulled up out in front of a rather posh hotel.

Joe paid the driver and craned his neck, taking in the opulence of the building before him. Never in his life had Joe seen a hotel of such splendor and, truth be told, it made him feel just a bit uneasy.

Walking up the front steps, Joe was startled when a uniformed man held the front door, bowing slightly while tipping his hat. Inside the hotel's grand lobby, marble tiled floors gleamed, reflecting the lights of an enormous chandelier suspended from the central ceiling.

Joe walked, spinning slowly taking it all in, toward the front desk where a rather stuffy looking man with thinning hair and a small graying mustache looked Joe up and down.

*"May I assist you Captain?"* the desk clerk inquired in such perfect King's English that Joe couldn't help but chuckle.

*"Is there some problem?"* the man inquired.

"No... Uh, no Sir!" Joe said, self-conscious now about his slight southern drawl. "I'd... I'd like a room..." glancing toward the piano player over by a fireplace, "... for a couple of nights?"

*"I'm terribly sorry Sir..."* and before he could finish his sentence, there came a clatter and loud laughter from the elevator bank facing the desk. Joe turn abruptly to see a trio of GIs, obviously well *lubricated,* escorting three ladies across the lobby. One soldier was trying to right a heavy brass ashtray he'd knocked over while the others hurried out the front door, the loudest of the group stuffing, what appeared to be a wad of bills into the breast pocket of the doorman's uniform with a salute.

*"...As I was saying."* The desk clerk reiterated, clearing his throat, *"I'm afraid we are currently filled to, or in some cases, <u>beyond</u> capacity."* shooting a gaze at the group that just exited.

"Well, that's okay." Joe said politely picking up his bag. "I guess a few fellers got here before me."

*"Yes, Indeed they have."*

## Robert Brun

Joe walked back out through the front door, which was promptly opened for him again by the same uniformed doorman. Somewhat embarrassed at his lack of finances, Joe handed the doorman the two coins he had in his pocket.

*"Thank yer sir."* the man replied with a bow and another tip of the hat that made Joe feel a bit like the High Roller he knew himself not to be.

Leaving the hotel somewhat faster than those around him, Joe was surprised to see the cab he'd arrived in still waiting at the curb, its driver leaning against its dented fender.

*"No room at the inn? 'Spected as much. As I said, been bringin' Yanks 'eer all day, I 'av. Need a lift?"*

"What I need is a suggestion." Joe replied as he tossed his bag into the cab discouraged, and got back in. "This place is the only hotel I know in London."

*"Well now, no worries mate. I know a place not 'af a mile from eer dat'll do you nicely for a cup-pul nights and at much bet-er rates at dat too."* and with that said, the driver pulled out into traffic, cutting off two MPs in an army jeep.

Not five minutes later they pulled up outside a small brownstone building right next to a bombed-out lot. The driver, opening the door, noticed Joe's gaze toward the bomb damage next door.

*"Don't you worry bout dat none. Jerry's not been back in deez ear parts for near on six months now, and it didn't touch dis place... septin a little."*

Grabbing Joe's duffle, the cabbie led him up a flight of stone steps and into the building that was small by comparison to the Palace, but looked clean and homey.

*"Di-ANN-a!"* The cabbie hollered as he set the bag down at the front desk.

*"Quit your shout-en, Reggie Baldwin or you'll be wake'n da baby... Oh!"* The woman said with a start, seeing Joe standing there.

An attractive redhead who appeared to be in her late twenties wiped her hands on her apron as she entered the hall. Suddenly looking embarrassed, she started to smooth her hair. Something about the way she moved reminded Joe of the day he'd brought the milk to the Fowler's farm, and he again felt a twinge of homesickness.

*"May I 'elp you?"* the woman inquired politely.

"Yes, I'd like a room for a few nights... Uh, one with a bath... perhaps?"

*"Well, la-di-dah!"* sang the cabbie tossing his head.

*"Reggie! Now you be 'oldin your tongue."* the woman scolded, as the man tipped his hat and headed back to his cab.

# My Shadow

"*Now never you mind about 'im.*" The woman said apologizing for the cabbie's behavior then looking a bit puzzled.

"*We'll get you set up just fine now. The room's not quite ready yet, but it'll be soon enough. Would you be 'ungry now? Dares a loverly pub just round da corner where ya kin git a bite whilst I fix up your room. Off you go now!*"

Without any say in the matter, Joe was whisked out of the front door as the woman picked up his kit bag and headed up a flight of stairs.

Actually, Joe was pretty hungry, so with no more than a smile and a shake of the head, he decided to check out the pub as suggested.

Walking down the front steps, Joe once again saw the cab driver waiting at the curb.

"*Send ya out for a bite did she now?*" the cabbie inquired as Joe descended to the street.

"Yeah, seemed in a bit of a hurry to get rid of me." Addressing the man, Joe added. "Listen, would you care to join me for lunch?" Joe asked the driver. I haven't had anything since breakfast back at the base and that was over seven hours ago."

"*Well... I don know?*" the man began hesitantly.

"Tell you what, instead of the fare, how bout I buy you lunch?" Joe added.

"*Well, what are ya waitin' for.*" and with that the driver set the parking brake and jumped out of the cab.

"*Reggie... Reggie Baldwin's me name.*" the man said pumping Joe's hand vigorously. "*Diana, dats me Sis.*"

Reggie spoke non-stop while the two proceeded down the block rounding the corner, a sign advertised The *'Keg and Flask- Public House'* that he assumed was the place Diana had recommended.

Joe and "*Reg,*" as he now wanted to be called, entered the pub, and it was soon obvious that Joe was not the first *Yank* Reg had brought here. Everyone seemed to know the man, and although they greeted him jovially enough, they seemed to keep their eye on him just the same.

The two men took a small table by the window near the front of the pub and soon the proprietor came over.

"*What'll ye be 'avin mate?*" the owner asked Joe, but before he could answer, Reg chimed in.

"*Two pints a Guinness and two boiled dinners.*" Reg ordered looking right at Joe.

"*And who'll be payin fer dis feast?*" The owner asked, looking right at Reggie.

"I'll be getting this." Joe said with a smile, enjoying the floor-show before him.

## Robert Brun

The pints arrived shortly after placing the order and he and Reg toasted the day. Looking around the room, several of the patrons also raised their glass to the young pilot. A bit uncomfortable, Joe squirmed in his seat, feeling less than the celebrity they were treating him as.

"Is it always like this in here?" Joe asked Reg who had already drained half his pint.

"Naw, they just have a thing for fighter pilots around deez parts. You see, we `ad a front row seat for the `Battle' back in `40. As far as dare all concerned" gesturing with his glass and spilling a bit on Joe's slacks, "all you flyboys is one of 'da few' as `Winny' would say."

"The `few?" Joe asked curiously.

"Yeah, you know:

`Never was so much owed by so many to so few.' Reggie said trying his best to imitate the Prime Minister, Winston Churchill.

"The ones dat stopped the Luftwaffe." and with that, Reg raised his glass.

Blushing slightly, Joe turned and raised his glass to the men there in the pub that nodded to him in return.

Three quarters of an hour later, now comfortably full and having bid Reg a `good day,' Joe returned to the brownstone and rang the bell in the front hall. Diana, now looking a bit more put together, greeted him with a big smile.

"Dare ya be now, did ya ave a good meal, take da edge off, did it?"

"Yes, it was delightful. Thank you." Joe said in reply.

"Well now, right dis way. I'll show ya to yer room."

Following her, Joe ascended the stairs and walked down a short hallway to an oak door. He took his kit bag from Diana while she unlocked the room and stepping aside, allowed him to precede her into the spacious room.

The room was truly elegant. A large bed filled most of it, and faced a tall wardrobe, flanked by two floor-to-ceiling windows with dark green velvet drapes held open by gold ropes. A large patterned rug covered the hardwood floor and atop the rug sat a carved oak desk and matching swivel chair.

"Dis eer is da main suite... an dis door..." As she lead the way to a second door beside the bed, "leads to the water closet."

Walking through the door, Joe found himself in a second room nearly as big and grand as the first. This room, however, was virtually empty except for a claw-footed bathtub sitting right in the middle. Off in one corner were a small sink, mirror and toilet, and through what was maybe the biggest window he had ever seen, Joe looked out across the rooftops of London. He turned to Diana and smiled in approval.

## My Shadow

*"Right! That'll be 3 and 6 a night. Breakfast is from 6 ta 8 a.m. We don't allow guest in da rooms and they'll be no noise after 9 p.m."*

"Uh, o-kay." Joe said slowly still admiring the accommodations. "That'll be fine." and took out his wallet.

*"You can pay me when you leave. Checkout is by 10 a.m."* And with that, she turned and left, shutting the door a bit harder than Joe thought necessary.

Joe, mildly stunned by the sudden change of character, tossed his kit onto the bed and watched it sink into the comforter.

"I think I'm gonna like it here."

Less than ten minutes later, he was easing himself down into a hot bath and staring out at the darkening London skyline. Lying back in the steaming water, he closed his eyes with a deep sigh.

It was a beautiful day as Joe flew the AT-6 "Texan" Advanced Trainer through the cloudless skies off the coast of Texas and out over the Gulf of Mexico. The open canopy let the warm, humid air of the Gulf blow through the cockpit and Joe could smell magnolias from the distant shoreline. This was the kind of day a pilot lived for, flying free with nowhere in particular to go and nothing to do, but log hours.

The "Texan" roared along effortlessly and Joe turned, banking the plane in slow weaving arcs. Flight training was finished and he had some time before heading off to fighter school. Just that morning, he'd received a letter from Mo that greatly bolstered his confidence in their relationship. Joe couldn't have felt better as he performed a perfectly executed Immelmann and headed back toward the coast. The bright orange summer sun had just begun to set over his left wingtip when the Focke-Wulf made its first pass.

Before it had a chance to register, the FW-190 opened fire stitching Joe's wing with 20 mm cannon rounds. Without so much as a though, Joe slammed the control stick off to the right putting the Texan into a hard bank. The plane pitched wildly over on its side while the FW-190 turned into him once again to begin a second pass. Joe took the initiative this time and lining up his attacker in the gun-sight, squeezed the trigger to activate the Texan's twin 30 caliber machine guns. Nothing happened!

A sound like hail on a corrugated tin roof resonated through the plane as the cannon shells of the Focke-Wulf struck home. The German fighter passed so close Joe could read the name *"Herr Doug"* painted across the enemy plane's cowling. The Texan's 600 hp. Wasp radial engine started to cough and oil spilled out from under the engine cowling as smoke and flames filling the cockpit. Blinded, Joe could feel the plane spinning out of control and knew he was going down fast. Trying desperately to unfasten his safety harness, Joe was slammed both by the impact and the cold water as the plane spiraled in, striking the Gulf hard. Feeling water entering his mouth and nose, Joe panicked, thrashing as he struggled for the surface.

With a jolt and a fit of coughing, Joe opened his eyes and sat bolt upright in the tub. The sun had set with the last glimmer of evening light streamed through the window forming geometric patterns across the walls and floor of the bathroom. The water in which he sat had cooled and the chill made him shiver.

Heaving a sigh and spitting out the taste of soap, Joe splashed his face with the now cold water and standing up grabbed the towel from the stool by the tub to dry off. Embarrassed by his actions, he mopped up the water around the tub and placed the wet towels over the radiator to dry.

Moving back into the bedroom, Joe switched on the reading light and glanced at his watch stunned to read 22:30. Feeling completely exhausted, he pulled down the covers and slid into the bed not bothering with his skivvies. Joe could not remember ever feeling so tired and comfortable at the same time, and soon he was sound asleep.

He awoke with a start at 05:30, thinking he had missed the mission briefing before remembering where he was. Feeling very hungry, it occurred to him he hadn't had any dinner the night before and wondered if it was still too early for breakfast. Getting up, he dressed and recalled the sitting room and fireplace across from the registration desk thinking he might wait there until breakfast was ready.

While shaving, Joe noticed a collection of feminine toiletries on the small table there by the sink. A few, barely noticeable dust rings from where items appeared to have been hastily removed were still visible. Joe just shook his head with a smile, realizing now why his hostess had been so keen to get rid of him for a while.

Back in the bedroom, Joe's curiosity got the better of him and opening the large wardrobe there in the room he saw several man's suits neatly hanging and pushed off to one side. He noticed several bore a monogrammed breast pocket with the initials H.G. framed inside a rather ornate family crest. Carefully he closed the wardrobe door.

Quietly descending the stairs, Joe was relieved to see that a peat fire was already blazing in the hearth, adding a warm glow to the otherwise darkened room, made even more so by a boarded up window at the far end.

There were several recent issues of the *London Times* scattered across the coffee table and picking up the most recent issue, Joe settled comfortably into a leather wing-backed chair and flipped through the paper, scanning the headlines.

## My Shadow

ALLIES ENTER ROME!
'RATIONING EXPECTED TO WORSEN.
'NEW GERMAN TERROR WEAPONS SUSPECTED,' and one that particularly caught Joe's eye:
'INVASION FEVER GRIPS NATION.'
*"Good mornin!"* came a pleasant voice from behind the chair. *"Did ya sleep well?"*

"Like a baby." Joe said standing with a smile as he watched Diana enter with a coffee pot and a tray of pastries.

*"Breakfast won't be for another 15 minutes, but I thought you might be likin' a bit of coffee prior. Or would you prefer tea?"*

"No, coffee will be fine. Thank you very kindly." Joe said taking the tray from the woman and setting it on the table.

"Joe." Joe said, turning toward Diana who at first looked at him a bit puzzled, then understood his outstretched hand.

*Diana... Diana Griffin."* the woman said, and wiping her hand on her apron, extended it to Joe. He took it and giving it a shake noticed the roughness of the skin. This was not a hand of the leisure class.

*"Reggie, that's me brudder. Ee 'elps me out by bringin me the occasional boarder. Times be-in what they is now, an all."*

"Yes, I can see that." Joe said gesturing toward the boarded up window at the far end of the sitting room.

Following his gaze, the woman stopped for a moment and sighed. *"Oh, that. Right. Came pretty close that one did. Back during the Blitz it was. Seems like donkey's years ago now. Luckier than most though, just took out the far window and..."* Her voice trailed off.

Joe stood silent for a moment.

*"Arold..."* There was a long pause before she spoke again. *"Went out to check on the blackout curtains, 'e did. Make sure they were set up all proper like...* There was another long pause.

*"...Well I best be getting breakfast ready for ya. How'd ya like your eggs? All I got is powdered, rationin bein what it is an all."* She said with a sniff.

"Scrambled will be fine," he said hoping he'd successfully hidden his disappointment while not knowing quite where to look.

*"Brilliant!"* she replied exiting the room.

The meal, a real *'fry-up,'* despite rationing and the SPAM - no getting away from it Joe guessed - was fit for a king. Joe enjoyed the company of Diana and her three-year-old son through most of the meal.

During their conversation, Diana told tales of the bombing and how much her life had changed in the last few years. She told about her son `Arry' and how he would never know his father or truly understand any of what this war was all about.

"I'm not sure I understand what this war is about myself, and I'm involved in it." Joe said, pouring himself and Diana another cup of coffee. "Of course, I haven't been though anything close to what you have."

*"Don't you go bein' modest now, Joe,"* Diana said, calling him by his first name, *"We may' ave been under da bombs, but no one was ever tryin to specifically kill me like you and de udder pilots, and besides,"* She continued while leaning across the table and gently touching his hand. *"Somebody has to stop dem."*

"Yeah, I guess," Joe said, sitting up and clearing his throat, " but it sure is disruptive to a guy's life."

*"Don't I know bout DAT!"* Diana said with a sigh.

"Oh, I'm terribly sorry!" Joe said feeling embarrassed by his thoughtlessness.

They sat there for a while, neither of them saying a word. Joe took another sip of his coffee feeling completely at ease. Surprised by how comfortable he was sitting there with Diana, Joe's mind started to wander and began to imagine the war being over. He saw the possibilities and hope began to grow inside him, picturing himself sitting in a kitchen just like this with a wife like Diana and a son just like little Harry. Looking across the table his eyes met Diana's. Though several years older, she was an attractive woman and Joe felt something stir inside him. It had been months since he had been alone with anyone like this and he wanted more than anything... he wanted... wanted...     Mo!

Like a bolt of lightning everything became clear to him and he knew it. Despite the lack of correspondence, the worrying and the waiting, he understood, it was Mo that he loved and Mo he wanted to be with when this was all over. It no longer mattered what happened, he would see her again, tell her all his feelings and make it all work somehow.

A loud **BANG** snapped Joe out of his fantasy and Diana turned to take a large wooden spoon from Harry who was now using it to keep time to some primitive rhythm on the kitchen floor. Diana looked as though she were about to say something, then stopped herself.

*"Oh, never you mind."* Diana replied,   *"Now finish up your coffee."* and got up to clear the dishes.

Following a relaxed breakfast, and after insisting he help with the dishes, Joe decided to take in some of the sights of London. The brownstone didn't appear to be too far from the center of town and, it being one of England's rare clear days, Joe decided to walk. Politely refusing the offer to call Reggie's cab, Diana pointed out the best route to get into the heart of London.

## My Shadow

Much of the area around the brownstone was fresh and clean, a 'leafy neighborhood' as the locals would say, and Joe enjoyed his leisurely stroll in the warm sunshine. As he turned another corner not much farther along, he encountered an area cordoned off with wooden barricades. In front of him stood what was left of a burned out residence, the ubiquitous UXB (un-exploded bomb) warning signs, clearly visible for the benefit of the curious.

Wandering further along the block with no particular destination in mind, Joe made his way through Islington and Westminster, Trafalgar Square and passed Nelson's column.

After grabbing a cup of tea and a scone at a quaint cafe, Joe found himself at the entrance of Kensington Garden where he noticed a familiar figure sitting on a park bench, drawing in a small book. It was Rob - with his cap down over his eyes shielding the sun, a second pencil held firmly in his teeth, and a look of intense concentration on his face. Not wanting to disturb him as he worked, Joe strolled around behind Rob and glanced over his shoulder at what he was working on.

On the pages of his book were several sketches of various recognizable buildings and structures from all around the city. He had been busy. After Rob put the final touches on the drawing of the Royal Albert Hall, Joe cleared his throat and Rob practically jumped out of his skin.

"Jeee-Eeez Georgia!" Rob blurted, whirling around and dropping his sketchbook. "Don't do that!"

"A good fighter pilot *always* watches his `six o'clock." Joe responded with a chuckle, surprised at the depth of Rob's concentration.

"Yeah, I know." Rob said contritely, "but I thought we were on leave."

"You're right." Joe said apologetically, sitting down on the bench beside him. "How's it going?"

"Great!" Rob replied, collecting his materials and handing Joe the sketchbook to look at.

Flipping through the pages, Joe saw drawings of the Tower of London, the Tower Bridge, Big Ben and several views of Parliament along with smaller studies of its various architectural details.

"Impressive." Joe said, handing the book back to Rob. "You have been busy."

"Yeah, I'm having a ball! All these places and so little time." Rob said, the excitement rising in his voice as he spoke. "Boy! I wish I had my paints with me. Just look at that sky, will you."

Joe glanced up at the cloud-filled sky over the park and immediately saw what Rob was talking about. He didn't need an artist's

## Robert Brun

eye to appreciate the white of the clouds against a cerulean blue sky which, with just a slight touch of pink, offset the brown/grey buildings beautifully.

"I'm heading over to the London Museum later this afternoon, want to join me?"

"Well, I don't know..." Joe started and a moment later they both heard it, a growing wail of what they soon realized was a siren.

"Air-raid!" Rob said snapping his book shut. "What now?"

Searching around the park for clues, Joe saw the people there only slightly quicken their pace while sirens continued to sound all over the city. Unlike those Londoners around them, this was the two pilots' first air raid, so Joe and Rob trotted over and began following the crowd.

Above the wail of the sirens was the low pulsing drone of the unsynchronized twin-engine aircraft followed by the higher pitched whine of a fighter and the familiar chatter of machine gun fire.

Glancing skyward, Joe caught sight of a German Junkers Ju-88. Black crosses and yellow wingtips visible, one engine smoking badly and with a lone Spitfire in pursuit.

The two aircraft passed overhead at barely 500' then both disappeared behind the skyline, followed by a low rumbling crash and the sound of tearing metal and breaking glass. Joe let out his breath and about five seconds later heard a loud explosion from a few blocks away.

Looking at one another, the two pilots took off running in the direction of the blast. As they ran, the Spitfire flashed overhead just above the rooftops, pulling up into a steep climb followed by a snap roll.

"Somebody's just made his day." Rob huffed as the two continued to run to the area where thick smoke was now rising.

Approaching the scene, several fire trucks raced passed, turning the corner at such speed that Joe was sure he was about to see the vehicles roll over in the turn.

The two men rounded the corner, coming to a stop and Joe gasped at the destruction he saw before him.

The bomber had crashed into a small building at the end of a street. The entire second floor was in flames from where the plane had impacted, its rudder sticking out through the conflagration, the charred swastika still visible on its tail.

Down below, a number of residents were running out of adjacent buildings coughing, handkerchiefs covering their nose and mouth's while the fire brigade set to work with the hoses extinguishing the flames fed from the bomber's ruptured fuel tanks.

Joe and Rob stood motionless, staring and listening to the popping sound of the bomber's machine-gun ammunition detonating in the heat.

"Move along... move along now. Nothing more ta see." a police officer called out, his arms spread wide herding the crowd of bystanders.

## My Shadow

"Is there anything we can do to help, officer?" Joe asked when the man was close enough to hear over the roar of the fire.

*"Tanks fo da offer, Captain, but we ave tings pretty well under control. Been through dis a few times before, we ave."*

"I suppose you have officer... I'm sorry."

*"Sorry....* The officer stopped for a moment looking back at the wreckage. *"Right! Now, move along."* and off he went shooing away the crowd of Londoners.

"Nothing more we can do here, Rob." Joe said tapping Rob on the shoulder, but Rob was frozen, his gaze locked toward the crash sight.

Looking back, Joe watched while the body of a young woman, not much older than twenty, was carefully lifted out of the rubble, her ripped skirt exposing an incongruent amount of leg. Her stockings were torn and one shoe was missing. An, expanding patch of dark red surrounded her right temple dampening her sandy blonde hair.

Joe watched the color drain from his wingman's face and who, seconds later, ran to the nearest lamppost and vomited.

# Robert Brun

## Chapter 16
## Leave London Part II

*Friday, June 2, 1944: A really rough day. Don't feel like talking or writing about it. Anxious to get back and get this dammed war over with! - R.B.*

The two pilots shared a cab, and after dropping Rob off at his flophouse - Rob hadn't had any luck at the *Regent* either - and refusing his offer to stay, Joe made his way back to the brownstone.

Diana met him when he entered the hallway and right away noticed something amiss.

"What's 'appened? Something's wrong, are you al-right?"

"Yeah" Joe responded, numbly. " I was witness to that air-raid downtown today."

"I heard. Was anyone hurt?"

"Hit a house over by Kensington. They pulled one girl out of the rubble. Dead!"

There was a long pause then Diana spoke up.

"Right... Now, you'll be-wanting your tea. Sit yerself down by the fire. I won't be five minutes." and before Joe could say another word she was off and into the kitchen.

Joe felt a strong ache in the pit of his stomach as he took a seat in the chair. The warmth from the fire felt good and took away some of the numbness he was feeling. He slumped down, staring at the flames, and tried to think it through.

In all the months he'd been flying, he'd never once given a thought to what it must be like below the fight... on either side. Truth be told, with all the fighting he'd been through, he'd never actually seen a casualty up

## My Shadow

close and this realization surprised him. His mood was not much brighter when Diana returned with the tea tray.

"There you be now, give 'er a few more minutes to steep and 'elp yourself to the biscuits."

As his hostess turned to go, Joe surprised himself by reaching out and grabbing her hand. She stopped and stared down at him.

"Please stay." he said looking up from the chair and a moment later releasing his hold. "Please?"

Diana let out her breath and after a pause, took the chair across from his. Sitting down slowly, she smoothed her dress and folded her hands in her lap.

"I saw someone killed today." was all that came through from the jumble of thoughts that now race through his head.

"I know. I've seen that look before. In me own mirror. When 'Arold was killed... I looked just like dat for months, "

"How'd you shake it?"

Diana's son ran into the room holding a toy Spitfire above his head. Dianna picked up the child and set the wriggling boy in her lap.

"You 'av to learn dat life goes on... whether we want it to or not. There was another pause and looking at her son.

"Ee's got is father's eyes and ee'll never even know the man ee's named for. A good man, 'Arold was, a good, kind man."

"But I'm part of it, I'm part of the killing and death, it just never occurred to me before today. They kill, I kill, what's the difference?"

Diana shot him a look of contempt, then relaxed her expression, looking more resolute then angry.

"Da difference ain't da _what_, it's da why! We didn't ask for dis war, but we couldn't just sit back and do nothing! Chamberlin tried, and look where dat got us. You keep on doing your part and bring an end to dis war as fast as you can because, as horrible as it all is, it's better to die fighting den to live as a slave."

Joe looked up for the first time since Diana had sat down, tears welling up in his eyes. He knew she was right, but the image of the dead girl was still too fresh in his mind. Lowering his gaze, Joe watched little Harry curl up in her lap gently stroking his mother's cheek, and the memory of Cushman flooded his memory once again.

Diana rose and touched his shoulder, paused for a moment and taking little Harry by the hand, left the room.

"Mommie, is that man sad?" he heard the boy say as they left.

"Yes, love, ee's very sad."

The next morning, Joe was awakened by the sound of loud knocking on the bedroom door. Opening his eyes and looking at his watch, he grabbed his trousers and headed for the door wondering what kind of emergency could warrant his attention at 04:10. Joe opened the door to see Diana standing there in her dressing gown, curlers in her hair.

"*Oh Joe, I'm terribly sorry to disturb you at dis hour, but dare's a mate O yours downstairs demanding to see you. Says it's important.*"

Joe grabbed his shirt and stumbled down the stairs to see Rob standing there pacing.

"What's up?" Joe asked, trying to clear the sleep from his head.

"Joe, I just got the word, all leaves have been cancelled. Everyone's been ordered back to the base. Pronto."

"The invasion?" Joe asked excitedly.

"Looks that way. Anyhow, I got a cab waiting outside ready to take us to the train as soon as you can get your gear.

Not ten minutes later, Joe had his duffle packed and was out on the front steps of the brownstone. Looking anxious, Diana was standing there with her arms wrapped tightly around her robe. Rob was down on the sidewalk pacing by the open cab door beckoning to his friend.

"*Take care of yourself.*" Diana said as Joe passed on his way out.

Pausing only for a moment, he smiled, leaned forward and kissed her cheek slipping a ten-pound note into her hand.

"For the accommodations." Joe said with a smile. "Thank you for everything."

Running down the steps, the two pilots jumped into the cab and headed back to Abington.

# My Shadow

## Chapter 17
## Day of Days

*Sunday, June 4, 1944: Back from London. All leaves cancelled. Looks like this is it, the invasion - R.B.*

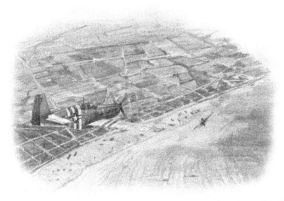

Catching a train back to Abington, Joe and Rob arrived to a flurry of activity. The base was locked up tighter than a drum and although everyone on base knew something was up, no one knew exactly what, other than that it was something BIG!

After stowing his gear back in his hut, Joe was still unsure what was going on. Lots of the pilots speculated, but no one know for sure so he headed over to maintenance. Zeke would know something, he always did. When Joe approached the hardstand where *MoJo* was parked, an unexpected sight greeted him.

"What the Hell?" Joe blurted out as he viewed his crew chief putting the finishing touches on the gaudiest set of black and white stripes that encircled his plane's wings and fuselage.

"Whad-ya think, Captain? Zeke replied with a smile, and pointedly ignoring Joe's angered expression. "They're all the rage!"

Speechless, Joe stood there when Rob strolled up alongside chewing on the last bits of a powdered egg and SPAM sandwich.

"Yeah, they hit *Shadow* as well. They're calling them *'Identification Stripes'* down at the motor pool and everyone's got 'em, though I must say, your guys sure did a much neater job than mine."

"But they're going to make us stand out like a sore thumb!" Joe said exasperated.

"That's the whole idea, Georgia. It appears we'll be flying over a good number of Army and Navy *Newbies* who've never even seen an enemy plane, much less fired at one. The stripes are supposed to help prevent us getting shot down by our own, trigger happy guys."

## Robert Brun

Joe stood there in disgust, then resigning himself to the situation, turned with a snort and stormed off toward the dispersal hut.

"What's up with him?" Zeke called down to Rob who was wiping his hands on his pants.

"I don't think he likes the new décor," Rob replied with a grin.

At the briefing the following morning, Joe and the rest of the pilots received their orders. Monday, June 5th, was going to be the start of the largest amphibious landing ever attempted, the invasion of the European continent. Thousands of ships and tens of thousands of troops would land along the coast of France at a place called Normandy.

For months, the allied high command had been maintaining a ruse that the invasion would be at Calais, the narrowest part of the English Channel. Reconnaissance photos had shown a massive build-up of enemy troops and equipment in that area. Apparently the ruse was working. Tomorrow, they would find out just how well.

The 597th's job was to fly close air support, escorting the bombers, and keeping the skies clear of enemy planes until a beachhead could be established and secured. The boys from the 364th and 479th Fighter Groups flying P-38s were going in first since it was believed that even the greenest gunners couldn't mistake the twin tailed P-38 Lightning for an enemy fighter. Joe and the rest of the 597th were to accompany the bombers, then fly inland and see to it that Jerry, whether in the air or on the ground, was kept away from the landing party.

Joe couldn't sleep that night, thinking about the next day's mission. Standing at the hut's window, he looked out into the darkness and listened to the enviable snoring of the other pilots. He kept wondering what it was going to be like for the soldiers hitting the beaches. Even from 15,000 feet, it was clear the fortifications the Germans had built for their *Atlantic Wall* were substantial. From what Joe had observed, Hitler's *Fortress Europe* was tight as a drum, and getting through on the ground was going to be...

...He didn't even want to think about it. Lighting a cigarette, Joe pulled on his flight jacket and cap and headed out into the night.

The warm damp air and heavy overcast made the night black as pitch as he made his way over to the hardstand where *MoJo* stood all tarted up in her new paint scheme. From the glow of his cigarette, Joe could just make out the five, foot-wide alternating black and white stripes adorning his plane.

"What have they done to you?" Joe asked the Mustang stoking it gently with his hand, the paint still tacky and the aroma of turpentine heavy in the thick air.

## My Shadow

With a sigh, Joe turned and noticed the faint orange glow of another cigarette a few yards away. Walking over, Joe saw his wingman leaning up against the side of his plane muttering softly.

"Couldn't sleep either?" Joe asked in a low voice.

"JEEZ! I thought I told you not to DO that!" Rob replied with a start that Joe had not expected from his usually calm wingman. "I'm already jumpy enough with all this secrecy."

"I guess we'll all know what's going on soon enough." Joe replied. You think this is really going to be it...? I mean the invasion?"

"I think so," Rob replied, lowering his voice and regaining his composure. "I've been listening to the dope, and if this ain't the real thing, I don't know what is."

The two men stood talking when a distinctive metallic `click' sounded from behind them, a sound that neither man had heard since basic training, but one that they both immediately recognized; the sound of an M-1 Garand rifle being cocked.

"Gentleman," came a familiar voice, "this is an excellent way to get yourself shot as Saboteurs."

Turning slowly around, Joe recognized the Corporal who had met them at the gate that night returning from the pub, but this time, he had his rifle leveled, steady as a rock, right at the two of them.

"I'm under strict orders to *'shoot to kill'* <u>anyone</u> outside and acting in a suspicious manner." the Corporal said with deadly seriousness. "And I'd say you two fit the bill quite nicely. Now, with all due respect Sirs, I'd suggest you both return to your quarters pronto, and stay there until further notice."

Rob started to say something in protest, but Joe, grabbing his arm replied, "Excellent advice, Corporal," and with a quick nod, Joe walked off into the darkness, his wingman in tow just as it started to drizzle.

The rain got heavier and around midnight, the word came down that the weather had put plans on hold for at least another 24 hours, the base, however, would remain on lockdown. No one could enter or leave.

Since, there wasn't much to do, Joe spent much of the day, like the rest of the pilots, hanging around the officer's club, drinking coffee, playing cards, reading, smoking and pacing the floor, alcohol consumption was out of the question. He'd tried several times to write Mo, tell her how he now felt and hope for some kind of confirmation, but each time he thought about the invasion and what it might bring. Maybe she had been right, not promising to wait for someone who might not return. Things here, he knew, were about to get a lot more dangerous. Tearing the half-written letter from the pad, Joe threw it into the stove and watch it burn. Then started pacing the room once again.

## Robert Brun

"Will you sit down, you're making my feet hurt!" Rob commented without looking up. "You're wound up tighter than my old man's mantle clock!"

"I'm just a bit anxious, that's all." Joe replied.

"Anxious! I'd say," came Rob's response, and he held up a sketch he had been doing of his Squadron Leader.

Joe glanced over at the drawing and was surprised to see a series of quick pencil studies that had perfectly captured the gestures of pure stress that he was now feeling. Heaving a loud sigh, Joe plopped himself into one of the battered armchairs.

"You're right," Joe sighed. "It's this dammed waiting that's getting to me. I hate it... just HATE it!"

"Well getting yourself all worked up over the weather isn't going to help, and besides, you're going to need your strength for when this finally does start. So sit down and relax, will ya?"

Joe paced the room, trying to calm himself, but with nothing to do and no orders except to stay put, it made for a very long day.

He occupied his time the best he could and finally managed to mail off a somewhat cryptic letter to Mo, knowing the censors wouldn't allow him to share any of his specific anxieties with her. Along with his words, Joe censored his feelings as well.

After sunset, when still no news had arrived, he made his way back to the hut to try and get some rest. The anticipation of the invasion allowed for little relaxation and Joe tossed and turned most of the night before finally managing to drift off to an uneasy sleep.

Later that same night and still well before sunrise, Joe was awakened by the sound of radial engines in the distance, what sounded like hundreds of planes flying over, and for the next four hours it continued non-stop. At 02:30 the following morning, the CQ roused Joe and the other pilots from their bunks with orders to be ready to fly in one hour.

Putting on his flight suit, still half asleep, Joe made his way to the briefing hut where he met the rest of the squadron, all looking as confused as he felt. After about five minutes, Colonel Tomlinson walked in, the men standing at attention, and headed right to the front of the room.

"At ease, men. As you have probably suspected, *Operation Overlord* has begun."

The room was instantly filled with wild cheering and shouts of "At ease...at ease." from the Colonel's aide.

"No doubt, you all heard the planes last night. That, gentleman, was the Airborne being sent on ahead to disrupt communications behind the German lines. A short time from now, our bombers will follow to soften things up a bit. Some of our fighter group, along with the Ninth Air Force, will be flying top cover for the bombers, and the rest will continue

## My Shadow

`nuisance bombing'` behind the enemy lines hitting anything that moves and creating general mayhem for the Germans. If everything goes according to plan, and there's been nothing so far that has indicated otherwise, we will have caught Jerry with his pants down around his ankles."

Taking a pointer from his aide, the Colonel stepped up to the map and pulled aside the curtain. Red yarn stretched from every fighter and bomber base in England converging at a place just east of the Cherbourg Peninsula. Joe noticed that some of the bombers would have to circle more than 100 miles west, out of their way before returning to form up. There was no doubt it was going to be a busy day.

"This, gentlemen, is where our bombers will be going and there'll be a *lot* of them. We're putting up everything we've got. We don't know for sure how much enemy fighter or flak activity they'll be, but the guys at S-2 seem to think it will be light. Orders are to concentrate our patrols in these areas," Tomlinson said indicating with the pointer to an area west of the beaches.

"And stay out of everyone's way. Basically, if you see anything on the ground or in the air and your not one hundred percent sure it's one of ours, destroy it! This is it men, the beginning of the end for the Germans and I wish like hell, I were going with you. Good luck, good hunting, and give 'em hell."

When the Colonel left the room, all the pilots stood at attention and starting somewhere in the back, someone began to clap. Soon the sound spread and Tomlinson paused only a moment before exiting to thunderous applause.

The men piled out of the briefing hut and into the pre-dawn darkness. After collecting his flight gear, Joe picked up his parachute, heading over to where *MoJo* was waiting.

"Well, this is it." Rob said approaching Joe from behind and slapping him on the back. "No more waiting."

"Thank God!" and Joe slung his parachute over his shoulder following the rest of the squadron out to their planes.

*MoJo* was sitting on the hardstand wearing her new paint job. Zeke had the engine running and was shouting something to the mechanic who was making some adjustments under the opened cowling.

"Anything wrong?" Joe inquired to his crew chief.

"Not a thing, Captain." shouted Zeke as he climbed out of the plane. "Just a few minor alterations that should give you just a bit more power."

"Thanks, Zeke, you're the best."

## Robert Brun

"You got that right, just bring my plane back in one piece and yourself as well." Zeke said, spat on the ground, and helped Joe strap into the plane.

"I'll do my best." Joe replied and settled himself into the seat scanning the instruments. Taking the creased photo of Mo out of his breast pocket, he once again placed it between the artificial horizon and the gunsight. Regardless of the lack of correspondence, she'd be flying with him just the same.

Following the directions from the ground crew, Joe eased the Mustang out from the hardstand and onto the taxiway. Moving forward into position, Joe looked over at his wingman who stuck his tongue out at Joe. Shaking his head with a smile and touching Mo's picture gently, Joe advanced the throttle.

Joe and Rob lifted off side-by-side and began their slow climbing turn to form up with the more than thirty-five other squadrons of Mustangs, Lightnings, Thunderbolts and even Spitfires that were patrolling the area over the beaches.

The slight cloud cover of that day partially obscured Joe's view below when they headed out over the English Channel along with more planes than Joe had ever seen flying at one time. Down below, Joe saw wave after wave of Fortresses, Liberators, Mitchells, Havocs and Marauders all wearing invasion stripes and all headed in the same direction. Beneath that, through the breaks in the under-cast, Joe got his first glimpse of the invasion force. The choppy water of the channel was so full of ships, their wakes drawing white lines in the dark water, that it looked as though it would be possible to actually walk across from England to France.

Closer to the coast, Joe could see muzzle flashes from the deck guns of the battlewagons and explosions on the beaches; Hitler's *'Atlantic Wall'* was taking a real pounding. Despite all of the activity below, there was no sign of flak or enemy fighters.

"We really must have caught them off guard this time." Joe thought, the French coastline passing beneath his plane.

"Hey Georgia, did you see that?" came Rob's excited voice over the radio. "Must be the entire allied army down there."

"Yeah, and from the looks of it, the navy too." replied Joe.

Beyond the beachhead, he could see the lines of hedgerows that made up the French countryside.

The four planes of Red Flight made their way inland with the evidence of the previous nights activities apparent below. Remnants of crashed Waco gliders were scattered over the fields, some of which were partially submerged in flooded marshes. Others had wings missing from impacts with posts that had been driven into the ground for just such a

## My Shadow

purpose. One particularly damaged Waco lay next to a line of bodies covered with a parachute. The *Wehrmacht*, even though caught off guard, was not going to make this easy.

When Joe neared the town of Pont-l'Eveque, villagers could be seen in the streets of the town waving white sheets and French flags as they flew over. Everywhere he looked there were planes all sporting the now familiar black and white stripes.

All day long, the pilots of the 597th flew back and forth across the Channel, escorting wave after wave of bombers, watching for enemy fighters and trying to avoid colliding with each other while the troops below fought their way onto the beaches. By 22:00, Joe and his wingman had flown their tenth of the more than 4,700 allied fighter sorties flown that first day.

Joe made his way back to the equipment hut and dropped his parachute on the floor with a thud. Sitting down on the bench, he pulled off his flight boots and overcome by fatigue, lay down along the bench with a groan. He had no idea how long he'd lain there before being disturbed by someone setting himself down beside him.

Opening one eye Joe saw the exhausted face of his wingman sitting there next to him, holding his head in his hands. His face was streaked with grime and a dark red ring encircled his mouth from the oxygen mask that had rubbed a raw spot across the bridge of his nose.

"Whew! What a day! What time is it anyway?" Rob asked gazing over at Joe.

Joe lifted his arm and studying his watch through blurred vision replied, "I got 22:35."

"22:35...Any chance the Colonel will let us sleep in tomorrow?"

Joe just shot Rob a look then closed his eyes.

"I thought not. Don't spend the whole night here, Georgia" Rob said with a sigh and getting up from the bench, left Joe lying there.

Hearing the door bang shut, Joe somehow found the strength to pull himself upright and make his way back to his hut. Without bothering to remove his flight suit, he collapsed onto his cot and soon was sound asleep.

Awakened the next morning around 03:00 to the sound of loud voices Joe soon discovered what all the excitement was about. Having, by now, figured out the ruse, the German Army was attempting to bring in troop reinforcements and armor under the cover of darkness. Joe and the rest of the Red Flight had been assigned to escort a flight of B-26 Marauders, a fast medium bomber, at first light, with orders to stop this advance movement.

## Robert Brun

Hurrying down a cup of coffee and two stale donuts, Red Flight was back in the air before dawn forming up with the Marauders.

Heading back out over the Channel for the eleventh time in two days, Joe was stunned to see the number of vessels now anchored right off the Normandy beaches. Several ships had been intentionally scuttled just offshore to create a breakwater or *Mulberry*, an artificial harbor to be used until one of the nearby French ports could be captured and secured.

Because of the heavy concentration of troops below, a *bomb-line* marking the boundary between the allied troops front line and the German fortifications was to be carefully observed, however, beyond this point, the Mustangs were instructed to attack anything that moved.

Escorting the bombers, Joe and the rest of Red Flight crossed over the French coast and were just inside Saint-Lo when they spotted a railway bridge spanning a small river near La Molay.

Giving the order, Joe had each member of the group assigned to a single bomber and led the four Marauders in single file from one thousand feet in a shallow diving run.

Releasing their bombs one after the other, the four B-26s made several direct hits completely destroying the bridge and collapsing it into the river.

Circling back around, Joe radioed the bombers and sent them on their way back across the Channel. Relieved of the burden of their charges, but with machine guns still fully loaded, Joe ordered Red Flight to split up and search the area for any enemy movement.

Joe watched Harrison and Taylor break off to the right leaving him and Rob heading east toward Caen.

"Mad-dog leader to Mad-dog two, come in Mad-dog two. Over"

"This is Mad-dog two. What's up Georgia?"

"Keep your eyes peeled for enemy movement down there and call out anything you see."

No sooner had Joe spoken these words, than a line of tracers shot up from a clearing below just missing his right wing. Rob and Joe broke left and right each circling around to join up 180 degrees from where they had been.

Joe keyed his mic. "Did you see where that came from? Over."

"Yeah, that clump of trees at two o'clock over by the farmhouse. It looks like a *Flakvierling 38* mounted on the back of a halftrack."

"A *Flak-er-what*?" Joe asked, surprised at his wingman's pronunciation.

"JEEEZE Georgia, don't you read the briefings posted from HQ? The German quad 20mm antiaircraft cannon! The one shooting at US!"

The line of tracers gave away the position of the weapon while the two pilots jinxed their way around the sky maneuvering into position for a

## My Shadow

shot, but each time they approached the clearing they were met by a solid wall of fire rising up from below and had to break off the attack.

"This is no good." Joe thought. Spending all his time dodging cannon shells he was unable to bring his guns to bear. Then Rob broke in.

"Hey Georgia, tuck in tight behind me and stay low. I know how we can get this guy."

"What are you up to this ti...?" and before he could finish, Rob was down on the deck racing along just above the treetops. The line of tracers followed the two planes as they descended, and just before reaching the clump of trees where the German quad hid, Rob pulled up hard, drawing the fire with him.

Like a light bulb going on, Joe instantly understood and while the tracer rounds climbed along with his wingman's plane, Joe saw the opening and triggered his guns walking the stream of bullets across the field and into the halftrack. Finding his mark, Joe concentrated his aim into the quad-20 ignoring the small arms fire from the troops below. Within seconds, the halftrack exploded and the tracer fire stopped.

Joe pulled *MoJo* skyward and formed up with his wingman who now had a gaping hole in the outer third of his port wing.

"That was close." came Rob's voice over the radio.

"And kind-a stupid? Wouldn't you say?"

"Hey, we got the guy, didn't we?"

Joe didn't reply.

"Head back to base." was all Joe said.

When the four Mustangs had landed, Joe was instantly over at Rob's plane shoving past Rob's crew chief, Tom.

"What the hell do you think you were doing back there?" Joe was furious.

"What do you mean?" Rob asked, genuinely confused by the question. "Knocking out a German antiaircraft battery!"

"Well, you could have been killed, you idiot!" Joe heard himself say before thinking.

"Georgia, my friend..." Rob said with more then a hint of sarcasm and placing an arm around Joe's shoulder, "We could all be killed."

# Robert Brun

## Chapter 18
## A New Ride

*Thursday, June 15, 1944: Man, am I beat! A whole week of non-stop flying. Georgia seems on edge lately. Maybe the stress of command is getting to him. I sure wouldn't want it - R.B.*

    For the next week, Joe and the 597th had little time for rest. Regular sorties back and forth across the Channel kept him and his wingman so busy, he'd barely had time to think about anything else.
    Nothing further had been said about Rob's flippant remark, and Joe knew Rob had been right. In this war there were no guarantees about <u>anything</u>! Anyone of them could be dead tomorrow and Joe began to wonder if his friendship with his wingman was beginning to cloud his judgment as a leader. As Squadron Commander it was his responsibility to order men into harms way.
    On D-Day plus 8, Joe returned from the day's mission, taxied *Mojo* over to the hardstand and opened the side canopy. *Mojo*'s Merlin had been acting up.
    "Zeke, could you check the sparkplugs? She's running a bit rough at cruise."
    "No need, Cap." came Zeke's off handed reply and spit on the ground. "I don't think it'll matter," gesturing with his thumb over toward the maintenance bay.
    Handing his flight helmet and oxygen mask to the crew chief, Joe started over toward the hanger where a gleaming new Mustang sat on the tarmac with the words *MoJo II* already adorned the cowling. The plane was pristine.
    Sun glinted off the polished aluminum as it sat in the afternoon light and Joe ran his hand lovingly over the smooth metal surface of the plane's starboard wing. The fighter looked radiant without the usual flat, olive drab paint of his B-model Mustang.
    "She's one of the new Ds," Zeke said coming up beside the Captain. "Just came in yesterday. She's 40 mph faster than the B, has six fifties, and the teardrop canopy will give you a full 360 degree field of view. I took the liberty of having Tom put on the name. Better take her

## My Shadow

for a spin, Cap, you'll be flying her into combat first thing tomorrow morning."

All the fatigue of that day's missions instantly vanished and like a kid on Christmas morning, Joe grabbed his helmet from Zeke and climbed into the new plane.

Settling into the cockpit, the smell of fresh paint and grease mixed with hundred-octane aviation gas filled his head making Joe slightly dizzy while Zeke helped strap him into the new plane.

"You only got about 30 minutes of daylight left, so better make it a quick one." Zeke said climbing down from the wing and signaling Joe to start the engine.

Advancing the throttle just a bit, Joe switched on the magnetos and battery. The instruments glowed under the panel lights and the gauge needles snapped into position with a barely audible *click*. Pressing the fuel booster and starter switch together, Joe watched the large four paddle-blade propeller begin rotating while the brand new Merlin turned over and caught in a cloud of white smoke. Like all the other Packard-Merlin's, this one too ran rough while the engine warmed up, but it was soon apparent that all 12 cylinders were firing properly.

Looking down at Zeke, Joe gave and received a thumbs-up from his crew chief, and, running the engine up to speed, checked once more to see that all was in order. Dropping back to fifteen hundred rpms, Zeke removed the chocks and Joe taxied *MoJo II* out along the perimeter track and onto the end of the runway.

Sitting there for a moment, Joe was reminded of the first night he'd barrowed the farms new pickup to take Mo to the high school dance. The smell along with the feel of powerful machinery in his hands, how similar it all was and how he wished Mo could be there with him now.

Looking down the field in the fading light, *MoJo II* cast a long shadow across the runway. Joe snapped back to reality, rolling close and securing the canopy. It was a strange feeling being surrounded by Perspex on all sides, and he couldn't help feeling a bit exposed.

Setting the brakes Joe advanced the throttle for the run-up then locking the tail wheel, released the brakes and the plane rolled easily forward. Once airborne, Joe activated the gear lever retracting the wheels and clearing the perimeter fence at the end of the runway.

Climbing out over the English countryside, he made note of the differences in the feel of this Mustang compared to his B-model. It had a few characteristics of its own, but overall, Joe knew there was nothing that he couldn't get used to. Being a test flight, Joe was able to relax a little and enjoyed just being able to fly again, something he hadn't been able to do since training.

## Robert Brun

That afternoon, while the sun sank low in the sky, Joe wasn't interested in altitude and kept *MoJo II* on the deck. Banking sharply to avoid a farmer's barn and scattering a flock of sheep across a meadow, Joe rocked the plane left and right just clearing stonewalls and stands of trees. Then, changing his mind he pulled back on the stick and pointing the plane skyward, pushed the throttle full forward to check the climb rate. The Mustang shot straight up, clawing for altitude as though unaffected by gravity.

At 10,000 feet, Joe retarded the throttle to idle until the weight of the plane brought him almost to a stop then, kicking hard on the rudder, yawed the plane over on its side and went into a steep dive. Watching his airspeed climb past 400 mph, Joe eased back on the controls and gently leveled off at 1,200 feet.

Joe gazed out across the wide-open English countryside forgetting about everything. He forgot about that day's missions, forgot about the bombers, the enemy fighters and the flak. The orange light in the western sky cast a warm glow over the panorama, *MoJo II's* wings gleaming gold in the sunset.

Reaching into the pocket of his flight jacket, Joe removed the creased photo of Mo and carefully slipped it under the new K-14A gyro computing gun-sight. How he missed her and wished he could share this joyous moment with her, tell her how much she meant to him. Flying was what he loved and, he had to admit, almost as much as he loved Mo, but he hated the fighting and the death. What he most wanted was simply for it to be over, to get on with life. A life somewhere quiet that didn't involve deflection shots, bomb placement and counting parachutes. A life where he could enjoy a good night's sleep and a decent meal, where days of boredom weren't followed by hours of anxiety and minutes of terror, a chance to find some sort of balance without the worries of the *Dougs* of the world. But he knew nothing was ever that easy.

It still bothered him that he hadn't heard from Mo for so long. How could she be so busy to not even write, unless...? This thought was interrupted by Rob's voice over the radio.

"Base to Mad-dog Leader, come in Georgia. Over."

"Base, this is Mad-dog Leader. What's up Rob? Over."

"Zeke wants to know when you're going to be back with his new toy? Over."

"Roger, making my turn now. ETA 10 minutes, Mad-dog out!"

Joe carefully removed Mo's photo and put it back in his pocket. Leveling off at 2,000 feet, he turned the Mustang on a reciprocal course toward Abington then noticed something cross the edge of his vision moving very fast.

## My Shadow

In the orange glow of the fading light, he could just make out what looked like a bright, pulsating tongue of flame streaking low along the southeastern horizon. Moving along at a high rate of speed, it appeared to be a small plane of some sort, but like nothing he'd ever seen before.

Banking onto an intercept course, Joe vectored the fighter to come up behind and to the right of the odd little aircraft avoiding the exhaust flame. He watched as the object flew passed making a strange, low buzzing sound audible even over the Merlin's roar. Then he noticed the white swastika outlined on the tail and all Joe's indecisiveness vanished.

Still not certain what he was dealing with; Joe sprang into action, calling on the instinctive behavior of his months in combat.

Hoping that Zeke had had the foresight to arm the new plane, Joe snapped on the gun switch and sight and tested the machine guns, relieved to see six streams of tracers appearing before him. Advancing the throttle to full power Joe settled in behind the strange little craft.

The thing was fast, but it held a steady course and at full throttle Joe was able to slowly creep up from behind it. Placing the pipper on the target, he could see the London skyline in the fading light. Time was running out and he was going to have to act fast!

Grasping the trigger on the control column, Joe fired at the object and immediately realized that, although the guns had been loaded and cocked, Zeke had not yet aligned them correctly - his bullets flew out wildly before him.

Frustrated by this new development, Joe lightly touched the rudder pedals and stick, rocking the plane back and forth and up and down trying to get at least a few hits on the target.

Concentrating on his maneuvering, flashes began appearing along the fuselage, his bullets hitting the small aircraft. Then, without the slightest warning, the small craft exploded in a huge fiery ball of smoke and flames.

## Robert Brun

"Holy Shee-it!" Joe cried in surprise, throwing the fighter into a hard right turn. There was noticeable *"thud"* from the left wing of *MoJo II* when he cleared the edge of the debris cloud and bounced onto his new heading.

Glancing back over his shoulder, the teardrop canopy afforded an unobstructed view of what was left of the downed aircraft, now burning harmlessly in a farmer's field.

Joe circled for one more pass and a closer look at the wrecked weapon before heading back to the base. He hoped the plane's gun camera had captured the event because there wasn't much left to see.

Rob and Zeke were both waiting when Joe taxied the shiny new P-51D back up to the hardstand. Shutting down the engine, Joe slid open the canopy and stuck his head out of the side of the cockpit.

"How was it?" Rob inquired, hopping onto the wing and helping Joe out of his safety harness.

"She's a bit heavier on the controls, probably because of the added weight of the extra 50s, but my guess is they'll come in handy. She's also a little twitchy on the yaw and the canopy will take some getting used to, feels a bit exposed, like you're sitting in a fishbowl, but the view is tremendous."

"Yeah, I can see that." was Rob's reply, looking around inside the cockpit. "Think I'll hang onto the *Shadow* for awhile longer. She's been good to me and I owe her."

"Suit yourself." said Joe jumping down off the wing and tossing his parachute to Rob who caught it with a grunt.

"Hey Cap, what the HELL happened up there?" Zeke asked staring at the large, ragged gash in the port wing's leading edge.

## My Shadow

"Uh... yeah. I had a little run in with something... I'm not sure what." Joe said, trying to ease his crew chief's obvious annoyance. "Kind of a... a flying bomb I'd guess by the looks of it... too small to be manned. I put a few rounds into it and ka-BLAM!"

"A V-1...!" Rob gasped, "You saw a *bloody* V-1 *Doodlebug*. I overheard one of the Spitfire pilots saying something about them at the pub the other night, all very top-secret you know. Operation *"Crossbow"* he called it. He said something about how one nearly killed him and a buddy in London last week. Took out a whole block of flats, flies by some kind of *radio beams\**."

"Well it must pack quite a punch!" Zeke added dryly, pulling at the torn piece of charred aluminum.

"Yeah, sorry about that. I guess I got a bit too close when it exploded."

"Yeah." Zeke said with a sigh crouching down under the wing to get a closer look at the damaged.

"This Brit says the best way to combat them is to come up alongside and *'flip'* em with a wing. Upsets their gyro. Too dangerous to shoot down." Rob added.

"Now you tell me!" and then turning to his crew chief, "Uh, and Zeke?" Joe asked reluctantly. "Could you see to it that the guns are aligned before tomorrow's mission?"

"Put on the coffee pot boys," Zeke said, to no one in particular, "It's going to be another long night."

\*Although at first believed to be true, it later turned out to be a completely mechanical guidance system.

# Robert Brun

## Chapter 19
## Letter from Home

*Friday, June 16, 1944: The new "Ds" are starting to arrive. Georgia got his and managed to dent the fender first time out on a V-1. If we didn't need pilots so badly, I think Zeke would have killed him - R.B.*

    Since before D-Day, the 597th had been flying close escort for bombing raids into Europe. The escort fighters, under orders to stay close to the bombers where they could be seen, helped with the morale of the bomber crews, but had proven to be an ineffective way of combating the enemy. The *Luftwaffe* fighters were best engaged out in front, well before they reached the bombers.
    When General James Doolittle, of the famed Tokyo Raiders, took command of the Air Force in Europe in January of 1944 he countermanded this standing order. Fighter pilots were now allowed to *"free hunt,"* encouraged to intercept and pursue enemy fighters away from the bombers, a decision that resulted in the destruction of a far greater number of enemy planes.
    The continued bombardment of German industry and cities by the USAAF and the RAF had also forced *Reichmarschall* Göring to recall his remaining fighters in defense of the *Fatherland* thus abandoning their forward airfields. With there relative short range, the Messerschmitts and Focke-Wulfs were no longer within range to defend the skies over France from their homeland bases. This left the area noticeably devoid of enemy fighters.

## My Shadow

However, in its continuing effort to disrupt the flow of German supplies and personnel to the ever-changing front, the role of the 597th over occupied territory was about to dramatically change.

It had been raining for almost three days straight when the letter arrived. It was the first Joe had heard from Mo for months. He hadn't had a chance to write since D-Day, and was anxious to hear any response to his letters so it was with both joy and trepidation that he opened the envelope.

*Dear Joseph,*
*I hope this letter finds you well and happy.*

There was something about the light-hearted feel of the greeting that made Joe apprehensive. He'd been through a rough couple of weeks and wasn't in the mood for pleasantries. What he really wanted was answers. He read on from there.

*Thank you so much for all your letters and the color photograph of you and the plane. I am truly honored to have it named for me. I think some of the other girls down at the plant are jealous.*
*Mother sends her greetings and I hope the cookies she sent made their way to you. I must apologize for the delay, but so much has been happening since last I wrote.*

Joe thought back to the box of chocolate chip *crumbs* he'd received months ago. He, Rob and the rest of the pilots had made short work of the whole, stale batch. At the time, Joe had thought it a bit strange to receive a package from Mrs. Fowler when he hadn't heard a word from Mo for so long.

*There has been so much going on and things here have been happening so fast I hardly know where to begin. The exciting news is I've been made Chief Fore-(wo)man of my section here at the foundry. Doug says that I am the best worker he has ever trained and keeps increasing my roll and responsibilities among his "Kittens." It's a big task and I hope I'm up for the challenge, but Doug keeps a close eye on me and spends extra time with me. He's very nice and has been so very supportive, yet still manages to find time to takes us out dancing once a week, although many of the girls are too tired after their shifts to go.*

*Doug* again! The name immediately jumped off the page.

## Robert Brun

*"Keeps a close eye,"* I'll just bet he does." Joe could feel his temper begin to flare.

*It's all so exciting helping out with the war effort... Not like you are of course, but in my own small way.*

*Well, anyway, there's the whistle ending lunch break and Doug is a real stickler for us "Kittens" being on time, so I got to go. I'll try to write again soon.*

*Your working girl,*
*Mo*

That's It!? No explanation for the delay. No reassuring words, just *she and Doug!* Joe crumpled the letter in his fist. What exactly was Mo saying? He tried to read between the lines and now wasn't sure of what he felt. *"Dear Joseph?* What had happened to *Dearest?"*

He reread the part about *Doug* for the forth time when something inside him snapped

"God-dam War!"

Joe picked up a water glass from the side table and sent it flying across the hut smashing against the doorframe just as Rob walked in.

"Whoa! The Orioles could sure use an arm like that. What's eating you *Mad-dog?*"

"Can it Rob!" Joe replied in a grunt. "I'm in no mood for jocularity right at the moment. What do you want anyway?"

"Colonel Tomlinson's requested all pilots and crews to report to the main hut at 18:30 today for a mission briefing." Ignoring Joe's venomous tone.

"A bit late for a mission briefing?"

"Didn't say what it's about, but orders are that anyone absent will be thrown in the brig." Then Rob noticed the torn envelope lying in Joe's bunk. "Letter from home?"

"Yeah," Joe spat and changed the subject, "Tell the Colonel I'll be there."

It was clear from the tone, that his Flight Leader wasn't interested in conversation, so Rob took the hint and exited the hut.

Joe looked at his wristwatch, 17:45, he had forty-five minutes before the briefing and he needed to pull himself together.

At 18:25, Joe entered the briefing hut and took a seat over by the open window. Despite the solid overcast, the day's heat had raised the inside of the hut to an uncomfortable temperature. This combined with the

## My Shadow

humidity and the smell of the other men in the room made it hard to breathe.

At exactly 18:30, the Colonel, along with his aide and Major Nealsen entered calling the men to attention. Joe retook his seat while Colonel Tomlinson made his way up to the platform appearing more reserved than usual.

Joe had seen the Colonel take this very same stage before every mission and always with a kind of enthusiasm that gave Joe and the other men the kind of reassurance they needed to complete each assigned task, but this time something was different and it showed. Reaching the center of the platform the Colonel cleared his throat to get the men's attention.

"Gentleman," the Colonel began, "For the past six months, the 597th has been escorting our bombers into the Reich and doing a dammed fine job of it. Men, I just received word from the top that there's been a change of orders. Staring now, and until further notice, our mission description has been redefined from bomber escort to *"Armed Reconnaissance/Ground Support."*

There was an audible groan from the assembly and Tomlinson remained silent while they did so. After almost a full minute, his aide called the men to order and the Colonel continued.

"As our troops push further and further into France, there is a increased need for our help at the frontline and that's what we'll be providing. The 9th Air Force has already gone on ahead and is making strong progress, but they need more planes. Our first A.R. mission begins at 05:30, day after tomorrow.

"Some pilots will be transferring over to the continent to be stationed on the ground with the troops as forward observers, CCAs or *Combat Command* observers. Wing Command believes that having men experienced in combat flying on the ground serving with the forward troops to call in the air strikes will do a better job, and with this, I agree. The assignments have been posted on the mission board. I know you'll do a good job and as always, good luck."

Tomlinson then turned abruptly and strode out of the hut.

The men sat stupefied for a few seconds then everyone started talking at once. Joe took the opportunity to avoid the crowd by exiting through the window he'd been sitting next to. Not sure how he felt about the change of orders and still pretty steamed about the letter from Mo, he lit a cigarette when Rob appeared at his side.

"I guess we say goodbye to the Big Friends for awhile. What about it Georgia?"

Joe turned slowly and looked Rob straight in the eye. "I think this war is about to get a lot more down and dirty." and as if to emphasize the point, he ground out his cigarette with the heel of his shoe.

# My Shadow

## Chapter 20
## Ground Attack

*Tuesday, June 20, 1944: Something's up with Georgia, not his usual self. Girl trouble I'll bet. New orders from HQ, seems we're gonna be down on the deck for awhile - R.B.*

  The P-51 Mustang is powered by the American built Packard-Merlin inline V-12 water-cooled engine. A marvel of engineering built under licensed authorization from the British Rolls Royce Company.
  It was this magnificent piece of machinery that replaced the underpowered Alison engine used in the first P-51As and A-36s, and made the Mustang the plane it had become by the summer of 1944. Its sleek profile and laminar flow wing along with the Merlin's 1,550 hp engine and 1,400-mile range, gave this fighter a distinct advantage over the P-38s and P-47s that had preceded it. As an escort fighter, the Mustang was outstanding, but in a ground-attack role, it was far more vulnerable than the air-cooled, eight gunned Thunderbolt.
  The P-47 Thunderbolt had been in service since 1942 and with its massive 2,000 hp. Pratt and Whitney radial engine, extensive armor and eight machine guns, was a real brute weighing in at over six tons. A true Juggernaut or *"Jug"* as it had come to be known. The Mustang, on the other hand, was a thoroughbred possessing a notable *Achilles Heel*. Having a liquid cooled engine, it carried its radiator beneath the fuselage, a very vulnerable location and one easily susceptible to ground fire. A single hole in the plane's exposed under belly and the plane would literally bleed to death. Without the precious glycol coolant that kept the Merlin's cylinders at the correct operating temperature, the engine would seize up within minutes.

## Robert Brun

It was an odd feeling lifting *MoJo II* off the airfield that first morning carrying the two 500 lb. bombs in lieu of the usual wing tanks, not that the added weight was noticeable. The bombs actually weighed less than the 206 gallons of extra fuel he usually carried on escort missions, but it was the addition of 1,000 lbs. of high explosives that Joe was unaccustomed to.

While at flight school, Joe had been trained in dive-bombing and had scored fairly well, but he never enjoyed it very much. There was something inherently wrong about diving a plane at such a steep angle toward the ground that just seemed to go against the very nature of flying. The escort missions he'd become accustomed too all were at altitudes well above 20,000 feet, he wasn't used to being so close to the ground and the thought unnerved him.

After forming up over the Channel and once again heading back to Cherbourg, Joe found his mind was still on the letter from Mo. He had poured his heart out to her and all he had gotten in return was the same evasive, non-committal response.

Scanning the instruments, Joe's eyes fell upon the photo of Mo beneath the gun-sight and he tried once again to interpret what she had written. What had she meant by `...best worker,'* and just what did this *'Doug'* guy mean to her? Joe's mind raced and spun until an agitated voice crackled in his earphones, violating radio silence.

"Mad-dog Leader, hold position!!!"

Looking up, *MoJo II* had crawled right on top of *Pretty Boy*, his Mustang having strayed out of its slot and almost into Wilson's plane. He was really letting this get to him.

"Snap out of it, Joe!" he thought, resigning himself to the situation, "There's nothing you can do about it now."

Taking a deep breath, Joe refocused his attention on the mission and these men who were depending on him. Turning his head, Joe looked over at his wingman who waved, made a *"V"* for victory and shot him the *'okay'* sign in rapid succession. Up ahead, the French coastline once again appeared out of the Channel mist.

Down below was the artificial port the Allies had constructed after securing the Normandy beachhead during those first bloody days. Joe observed columns of American and British trucks and tanks working their way inland from the beachhead and noted the evidence of the fighting that had taken place there was visible even from above. Further along, he saw line after line of the hedgerows that divided the French countryside and had made the fighting so rough for the ground troops.

The four planes of Red Flight continued inland and flew north along the route of the German retreat. Lt. Harrison was away on leave and

## My Shadow

Taylor was down with a sinus infection so Lt. Ben Wilson and Lt. Bud Smith were making up the remainder of Red Flight.

After less than ten minutes, Joe spotted what they had been looking for. Off in the distance he could see a column of smoke that piqued his interest.

"Mad-dog Leader to Red Three and Four. Wilson, you and Smith break right and head over toward that smoke column. Rob, you stick with me and we'll cover sector fourteen. Over."

Joe heard three rapid *clicks* indicating everyone heard and understood as two of the Mustangs broke right and headed away in a retreating arc.

Rob pulled in close to Joe's left wing, *MoJo II* and *Shadow* continuing on their course. They hadn't traveled far before finding what they had been searching for.

Up ahead along a row of trees was a road jammed with retreating German vehicles. Trucks, troops and horse drawn wagons were lined up end-to-end and stretching out for almost a mile. The two Mustangs approached the road, lining up their planes with the column of vehicles.

The sound of the *Jagdbombers* or *Jabos* - as the German's called the fighter/bombers - had alerted the soldiers below to their presence and men began to scatter into the trees. A few more dedicated soldiers took up their rifles, firing at the two approaching planes.

Joe leveled off parallel to the hedgerows on either side of the convoy 350 yards ahead. Ignoring the ground fire, he throttled back slightly and pulled the trigger on the control stick activating the six fifty caliber machine guns.

Bright, white-hot tracers shot out in front of him kicking up small puffs of dirt along the surface of the road. Adjusted the controls slightly Joe brought the nose of his Mustang up until his bullets found the target. The last truck in the convoy exploded throwing burning, flailing bodies in every direction. More soldiers ran while Joe applied slight right and left rudder slowly moving his fire from side to side while proceeding along the line of vehicles and men.

Despite his seven and a half months of aerial combat, nothing he had seen so far had prepared him for what he was now witnessing ... No, causing! He could see his bullets mowing down men, horses and machinery in a way he had never witnessed when engaged in aerial combat. Metal, uniforms, and flesh were being ripped and shredded all along the length of the road in apparent slow motion. Several of the trucks and half-tracks were now burning along with their unfortunate occupants. Horses panicked and reared up overturning wagons filled with troops only to fall again writhing and kicking in agony.

## Robert Brun

Joe centered his plane over the road and a strange feeling of detachment came over him when he pulled the twin levers that released his two delayed fuse 500 lb. bombs. The fighter rocked and lifted noticeably with the weight of his bombs falling away.

Pulling up slightly, his plane reached the front end of the convoy and Joe turned to see a huge cloud of black smoke rising up from the carnage he had just created.

His mind numbed at the realization of what he had just done, the men he had killed and wounded, and this time there was no kidding himself that a pilot had bailed out or managed to safely belly-land his damaged aircraft. These men were dead and he had caused it. He felt sick.

"Red two to Mad-dog leader, eggs in the pickle barrel. Do we go around again? Over." came Rob's inquiry over the radio.

"No," Joe said curtly. "Rendezvous with Red three and four and head back to base. Out!"

Without another word, Rob tucked in close to his flight leader and followed Joe in silence.

Later that evening, sitting by himself in a far corner of the crowded Officers' Club, Joe was feeling miserable. With the almost non-stop flying they had been doing since the invasion, many of the men were blowing off steam and over at the bar, the US Army Air Corps provided liquor was doing its job.

Joe usually didn't imbibe the hard stuff, leaving that type of drunken revelry to those with livers better suited to the task, but tonight, he felt differently. Tonight he'd purchased an entire bottle of whiskey and had already consumed more than half of its contents... alone.

Despite the festive activity surrounding him, his first day of this new assignment was weighing heavily on his mind. Consciously, he knew this too was part of the job he had signed up for, but was this really the way it was going to be from now on, and how long would that be? Each time he closed his eyes, he could see German soldiers running from the convoy he'd savaged. The terror on their faces as he flew over was clear in his mind and he knew that he had been responsible.

Never before had Joe considered what it must be like to feel the impact of a fifty-caliber slug. Today he had seen the results with his own eyes. Try as he may, he couldn't shake the images of grey uniforms and red blood splattered across the surface of the vehicles he had strafed.

Joe lowered his hand from his forehead and felt his arm bump the glass knocking the tumbler of whiskey off the side table and onto the cement floor where it shattered.

## My Shadow

"Dammit!" Joe shouted and downing a large gulp straight from the bottle, stood ignoring the crunching of glass under his feet and exited the club, bottle in hand.

The night air was cool and damp and he weaved his way across the field onto the apron where his plane stood patiently. In the dim light of the quarter moon Joe could just make out *MoJo II* painted across the engine cowling.

Seeing this, he felt ashamed that he had put Maureen's name on this murderous weapon, this instrument of death that it and he had become. Stumbling backwards and raising his arm, he threw the empty bottle at the Mustang, missing it by several feet and heard it shatter across the pavement of the hardstand. Then leaning back against the sandbags that surrounded the plane, Joe lit a cigarette allowing his feelings of self-pity to embrace him. He blew out a cloud of smoke and let the weight of his body pull him down along the sandbags where he passed out.

Joe didn't know how long he had been sitting there, but it was still dark when he opened his eyes again. His uniform trousers were soaked through and the inside of his mouth felt like a mixture of aviation fuel and peanut butter. With a tremendous force of will, he managed to drag himself to his feet and look around. *MoJo II* was still standing there and walking over, he rested his face against the smooth surface of the bare aluminum fuselage. The coolness of the metal felt good against his cheek.

"I guess we'll do what we have to." he said as a light drizzle started to fall.

"There you are!" came a voice from out of the darkness. "I've been wondering where you'd gotten off to."

Joe made a wobbly turn and looked around to see his wingman coming toward him following a flashlight beam.

"I've been looking for you for over an hour. The guys at the OC said you took off in quite a huff."

"What the *hell* do you want?" Joe found himself saying.

"Whew!" said Rob, waving a hand in front of his face. "You smell like the men's room at the pub. Better stay away from that stuff or you'll end up like those bomber boys."

"Yeah, and what about it?" Joe snapped in a slurred speech, "Maybe they all got the right idea. Maybe they're the ones doing the REAL work here. Blowing those Nazi cities to kingdom come, and killing everyone of em! Besides, what's it to you anyway, Lieutenant *Swiss Cheeeeeeese*?"

Rob stopped and stood for a moment without moving, his face illuminated by the eerie glow of the flashlight's reflection off *MoJo II's* aluminum skin. Looking both angry and hurt, Rob stared while Joe swayed slightly in the flashlight's beam.

"That one you get for free, *Captain*," Rob said taking a deep breath and, trying to control his anger, "The next one might just cost you plenty."

For the next few seconds, the two men stood toe-to-toe staring at each other, not saying a word. By now, the drizzle had become a steady rain and both men were soaked through. Then Joe's balance wavered, and he stumbled ending the stand off.

Rob let out a deep sigh and shook his head. "Come on, let's get you to bed." He put his arm under his Squadron Leader's and the two hobbled their way back across the field to the officer's quarters. Rob had just opened the door to the hut when Joe spoke again.

"How do you do it?" Joe blurted out when they were inside.

"Do what?" Rob asked confused.

"Stay so calm... so cool, so collected? Never bothered by *anything*. You were there today, you saw what I did!"

"Yeah, I was there and I saw what *WE* did... You forget two of those bombs were mine? " Rob let his voice trail off. "What makes you think it's any easier for me?"

Joe made no reply while Rob guided him over and sat him down on his cot. Taylor and Wilson were already asleep and snoring loudly. Rob glanced over at Harrison's empty cot, the one where, just a short time ago, Cushman had slept.

"You really want to end up like those guys?" Rob asked as Joe sat on the bed struggling to remove his wet uniform.

"After every mission, they drink themselves half-blind and pass out in their bunks. What kind of life do you think they'll have after this is all over? I think Mo deserves something a bit better then that, don't you?"

"Yeah... Mo!" Joe scoffed, staring up at Rob.

"Now get some sleep Georgia. We're flying again in..." and glancing at his watch, "...four hours.

Joe closed his eyes and fell back onto the pillow while Rob tossed the blanket over his friend's motionless body.

Exiting the building, Rob closed the door of the Nissen hut behind him. Rounding the corner, he pulled the last cigarette out of his third pack that day and putting it between his lips, crushed the empty pack in his fist. At the same moment, Rob's whole body began trembling so much that he had to steady his Zippo with both hands just to light it.

"Calm, cool and collected," Rob thought, snapping the lighter shut, "Yep, that's me, in spades.

### My Shadow

# Part III
### Chapter 21
### Transfer

*Monday, July 29, 1944: Had a run in with Georgia last night. Man, I've never seen him so drunk, but after the day we had, can't blame him. I wish it were that easy for me - R.B.*

Ever since the 597th Fighter Group's orders had been changed from bomber escort to ground support rumors been circulating about a possible transfer of operations from Abington over to the Continent. The Allied Forces, having established a foothold in Normandy, were fighting to hold on to what had been so hard won.

At the same time, the Germans, having realized their mistake were making every effort to bring in reinforcements. In order to prevent this from happening, the 8th and 9th Air Forces had to maintain air superiority while at the same time halting enemy movement and destroying all their equipment on the ground.

The 9th Air Force had already been moved to the Continent and stories of frontline living conditions at these new forward bases in France were rampant. These rumors were met with both increasing disdain and concern. Canvas tents sitting in the mud over wooden slat floors. Open holes for latrines, bad water, worse chow and rats and other vermin all made Abington's sparse accommodations now seem like paradise.

On the morning of July 15th, Joe and the rest of the 597th got word of a special briefing called at 09:30. Colonel Tomlinson would be settling the matter and putting all rumors to rest once and for all.

The Colonel strode to the front of the room with an expression that gave nothing away. In addition to the pilots, many of the ground

## Robert Brun

crews had also squeezed into the back of the hut, some leaning through open windows. The briefing was standing room only.

"Gentleman," Tomlinson began. "As you know, the 597th's mission for the last year has been escorting our heavies on missions to and from the Germany. However, since the Normandy breakout that has changed and now, so has our base of operation." There was a pause, and then he continued.

"There has been a great deal of talk lately about the next step and what that means to the 597th and I'm here to set the record straight. As of today, most of you will return to escorting the bombers back into Germany, however, some of the group will continue on with our present *Armed Reconnaissance* role, but will being operating from a forward airfield on the Continent." There were audible groans.

"The names of the men being transferred have been posted on the mission board and a briefing at 11:00 is to follow. To these men, I say, good luck and God speed. Dismissed."

Looking disgusted, Tomlinson strode out of the briefing.

Air Landing Ground 7A (A.L.G. A-7 for short) had been in operation since the 28th of June at Fontenay-sur-Mer and was nothing more than a bulldozed apple orchard covered over with Marsden Matting.

Reports were that these forward bases were so close to the front lines that in addition to the discomfort of sleeping in tents and eating K-rations, the pilots also had to deal with enemy artillery fire as well. After a momentary lull at the briefing, there was a mad rush to the mission board to check for names. The swearing and cheers that followed were a good indication of the general feelings from the pilots regarding their new assignment postings.

Joe swore when he saw his name on the transfer list, but with a further scan, felt both relieved and at the same time a slight guilt to see Rob's name listed along with the rest of Red Flight.

Outside the hut, Joe found Rob speaking with Taylor and Harrison about the pending move.

"Well, it looks like we're all the lucky ones." Joe said making note of the sour expressions.

"Lambs to the slaughter's more like it." Taylor intoned. "Have you been following the scuttlebutt from over there? Tents... they'll have us living in tents... in open fields! Boy I never knew how good we had it in the Nissen's. That bastard Tomlinson! He must have some kind of grudge against us! What did we ever do to him, huh? We fly his dammed missions. Pulled the Big Friends fat outa the fire more times than I can count and what do we get for it? Transferred to some stinkin *Frog* hell-hole"

## My Shadow

"We's just goes where they sends us." Rob chimed in a singsong fashion. "Actually, I'm kinda looking forward to it, I ain't never been to France." and then he added, "On the ground, that is."

Taylor punched him in the arm and Rob winced.

"Shut your trap Browning." Taylor hissed through clenched teeth. "You and your ga-dam optimism make me sick sometimes."

"All right men," Joe interrupted, but Taylor went on.

"I've just about had enough of this A.R. bullshit, I came here to shoot down German fighters, not wet nurse the dammed infantry."

"That's ENOUGH Lieutenant!" Joe fixed his stare on Taylor. "Get your gear together. Like it or not, we're shipping out."

Two days later, the four pilots, along with eight other Mustangs from the 597th loaded their personal belongings into empty, modified drop tanks the maintenance crews had prepared. Hatches in the wing tanks provided the pilots extra storage to carry the equipment they would need immediately on the short flight across the Channel. Joe stood on the wing of *MoJo II* stuffing the last of his gear behind the cockpit seat when the Colonel's staff car pulled up along side. Joe came to attention and saluted.

"At ease Captain. I can't say I'm pleased about breaking up the group this way, but I know you men will make the 597th proud."

"Thank you sir. We all follow orders. The men and I will do our very best."

"I know you will. You'll be the senior officer there, so I'm counting on you to look after the men." There was a pause. "One other thing Captain, I've been in contact with the CO of operations over there and he's assured me that you men will be well looked after. Good luck Joe." The Colonel saluted and with a trace of a smile got back into the staff car, driving off. Joe finished stowing his gear.

Following the C-47 transport sent back to escort them over and carrying the supplies for the new base, Red Flight made the short hop over the Channel and into France. Joe could see the A.L.G.s at Maupertus and Azeville on the tip of the Cherbourg Peninsula, but it quickly became clear that this was not to be their final destination.

When their flight approached the designated coordinates, Joe was surprised to see, not the gouged out landscape of a hastily assembled A.L.G. but a proper airbase complete with multiple, intersecting concrete runways and support buildings. This all appeared to be a step up from Abington's grass strip.

The last of the five planes to touch down, Joe taxied his fighter onto the hardstand and was very surprised to see Zeke sitting behind the wheel of a jeep, cigar held between his teeth.

## Robert Brun

"Zeke!" Joe shouted above the growl of the engine. "How the hell did you get here?"

"Came over on one of the *Dakotas* yesterday laying on top all the crates. Pretty tight quarters, but it sure beat going by boat. Besides, ain't nobody else gonna be looking after Red Flight's *SPAM Cans* `cept fer me." Then, without missing a beat. "Cap, have you seen this place? It rivals gadam Buckingham Palace! An abandoned *Luftwaffe* airfield! Boy, I tell you, those *Boche* bastards sure know how to live. Smashed things up pretty good when our guys asked `em to leave, but most of that was just cosmetic." Then excitedly he added,

"There's an in-tact FW-190 over in the #2 hanger I can't wait to get a look at. Hoo-Wee!" he cheered, "I'm in Hog-heaven." and removing his cigar from his mouth, spat on the ground.

Hoisting his parachute, Joe made his way passed groups of unfamiliar GIs milling around and poking through things looking more like tourists then occupying troops. The 597th hadn't been the only squadron tapped for this duty, several other 8th Air Force squadrons had contributed planes and ground crews as well.

"Joe, over here!" Rob called coming out from the airfield's main building, a 19th century French château.

"These," Rob announced with a grand gesture, "are our quarters." and with that, he pulled open the heavy carved oak doors to what had once been a grand mansion.

"This place used to belong to one of France's wealthiest vintners before the *Brats* barrowed it back in `40. Judging from the looks of the place, I'd say they left in quite a hurry."

Joe noticed the table in the main dinning room was still set for dinner. Fine china and silverware flanked by cut crystal goblet half filled with red wine. Flies buzzed around the spoiled food while the mess crew was busy clearing up.

"Your room is at the top of the staircase. Don't mind the patched hole in the ceiling, the *Brats* needed a little *persuading* before they agreed to leave, if you catch my drift?"

Climbing to the top of the stairs, Joe entered a spacious boudoir painted in several shades of pink and white with gold leaf adorning the relief trim along the ornate walls and ceiling. He felt as if he had just entered the bedroom of some fairytale princess and his first impulse was to wipe his feet.

In one corner of the room was a pile of broken plaster below the hole Rob had mentioned. It looked oddly out of place in such an otherwise pristine room. Shafts of sunlight streamed through the boards and a canvas tarp that had been hastily nailed over the opening to keep out any rain. Fortunately the summer nights in France were mild.

## My Shadow

Feeling more than a little out of place, Joe tossed his jacket on the bed and sat down in one of the pink upholstered chairs.

"I guess this'll have to do," he chuckled.

There came a knock at the door and Rob sauntered in carrying a bottle of wine in each hand.

*"Rot oder weiss?"*

Joe just sat there looking puzzled by his wingman's German.

"Red or white?" Rob asked again, this time in English holding out the bottles. "There must be hundreds more in the cellar. Spoils of war and you should see the champagne!"

"Not too bad." Joe said thinking back to what Colonel Tomlinson had said about seeing to it his men would be looked after. "Not bad at all."

That evening, after an incongruous dinner of warmed K-rations accompanied by fine wine in crystal glasses served on fine china in the opulent dining room on the main floor, Joe excused himself and went up to his room. Seated over by the fireplace that now glowed a deep red, taking the chill out of the air, Joe kicked off his boots and removed Mo's letter from his pocket. Rereading it for the tenth time, the same words slashed out at him: `Doug, dinner, dancing, his kittens!'` and the same anxious, emotions began again. Like waiting for the invasion, Joe's inability to know or do anything bugged the hell out of him, but he still knew there was nothing he could do. In the meantime, he had to get on with his duties and hope for a swift end so he could get back to her. He refolded the letter and put it away.

There was a gentle knock at the door and Lieutenant Taylor entered looking a bit sheepish.

"What is it Lieutenant?" Joe asked.

"Captain, the men and me have been talking and we think...?" his voice trailed off. Joe sat down, absorbing the pause and letting Taylor take his time.

"...We think... That is to say, I think that maybe I shouldn't have shot my mouth off back at the base the other day. I mean, it's not your fault that we got transferred, that's all up to Tomlinson and he has to follow orders same as us. It was just that I... I was pretty steamed after hearing all those rumors about the duty here on the continent and all and well... I wanted to say I was out of line... I mean LOOK at this place will ya! It's a ga-dam palace! ...Sir!"

Joe sat in silence looking up at the Lieutenant. It was clear that Taylor's unease was growing into embarrassment. Whatever he might think of their CO, Colonel Tomlinson did look after his men. Just how he'd managed to pull this one off though, was a true mystery.

# Robert Brun

He also had to hand it to Taylor. His loud-mouthed bravado never let on that he was even capable of such self-recrimination. Joe was proud of him, but also knew it would be wrong to say so to a man like Taylor.

"I'll see to it that the Colonel is informed that you approve of the accommodations, Lieutenant." Joe responded with more than a trace of irony. "Now get some rest, we have a busy day tomorrow. Dismissed."

After Taylor left, Joe walked over to the fireplace and lifting the poker, jabbed at the glowing embers. The room was warm, his men were comfortable and for the first time since arriving in theater, Joe felt like he knew what he was doing.

# My Shadow

### Chapter 22
### Tiger, tiger, burning bright

*Saturday, September 16, 1944: Getting settled here in "Paradise," France. Despite our dirty job, at least the accommodations are good - R.B.*

    The Merlin droned on while Joe and Rob completed the forth circuit of their assigned sector over northern France. For several hours the two had been flying a scouting AR, searching for potential targets and had yet to see another plane, much less any German troops. The fighter sweeps that had been conducted since the Normandy invasion had been effective, nothing was moving. The allies still maintained control of the air and from their forward base in France, the 597th, along with those of the other fighter groups of the 8th and 9th Air Forces had pushed the *Wehrmacht* out of the area as well. It was an unpleasant duty, but after two months, Joe had more or less, become accustomed to the routine.
    There had still been no word from Mo, but mail wasn't the only thing slow in getting to their new base. Spare parts had been in short supply since the transfer and fighting at such a low level, aircraft bomb damage was common, but Zeke and his crews had been able to take up the slack... so far.
    "Mad-dog Red Flight Two, this is Mad-dog Leader. Come in. Over." Joe called to Rob over the radio trying to ignore the muscle cramp in his shoulder.
    Red Flight Leader, this is Red Two, what's up Georgia?"
    "All seems pretty quiet to me, Red Two. What do you say we make one more pass through this sector and head back to base?"
    Joe heard Rob key his mic once to indicate his agreement and the two Mustangs made a slow left turn back to the east for a final pass.

## Robert Brun

Joe began another turn over the expansive farmland where everything was peaceful. Below, cattle stood grazing and acres of apple orchards and vineyard spread out as far as he could see. Framers driving teams of horses plowed fields while villagers rode bicycles along dirt roads. The bucolic scene was a far cry from the raging battlefield the area had been a few short months ago. The scene gave Joe a twinge of homesickness and he wondered once again what was going on with Mo back in Georgia. He was afraid to write after her last letter, but had hoped to hear something soon. Stationed in France for two months now and still there had been nothing, not a single word from home.

Joe set his left elbow on the armrest and shifted in his seat. The canvas cover of his parachute and inflatable dinghy made for a hard cushion against his backside. What he wouldn't give for a nice overstuffed armchair right now, but then no armchair could carry him safely back to earth out of a burning plane or keep him afloat in the frigid waters of the North Sea. He could deal with the discomfort for a while longer, besides, the way things were going, he'd be back at the base soon enough for his aching behind.

Executing a banking climb, Joe leveled off at 2,500 feet when his wingman pulled into position beside him.

"Hey, Georgia, Joe heard Rob's voice come over the radio. "I'm picking up a broken call on B-channel."

Joe tuned in his radio to the Air Support frequency picking up the broken call. It sounded like one of the pilot air controllers stationed with a forward armored division. The CCA's job was to observe and call in air strikes when and where needed. A pilot's understanding of the logistics involved in these air strikes made him much more valuable to the AR planes flying above. The desperation in the man's voice was evident even through the heavy static.

"...rcraft in... Any ...rcraft in ... area, this is for...rd obser... 23rd infant... armor division. ...Under atta, repeat, under attack ...multiple Tigers ...need of immediate assist... Do ...copy? Over."

"Rob, did you hear that?"

"Yeah, sounds like Rowan. Wasn't he assigned to the 23rd?"

Captain James Rowan, still recovering from the wound he'd received back in March had been assigned to the front line with the 23rd Armored Division. His job was directing Close Air Support for the troops making their way across occupied France.

"CCA air controller, this is Red Flight Leader Mad-dog, Shadow, Mad-dog. What is your position, over."

"Mad-dog... Is that you Joe? This is Rowan. We're five miles south, sou...east of Vire in sector 147. We ...under attack by two Tiger ...anks and are need of assistance, PDQ. Do you read, ...ver?"

## My Shadow

Joe checked the sector map attached to his right thigh and located the coordinate area. Switching back to the fighter channel, Joe called over to his wingman.

"Rob, how are you set on fuel? Over."

"Still showing enough for a few passes and I haven't fired a round yet, but what do you expect to be able to accomplish against Tigers? Over."

"I have no idea." Joe replied. This had been a scouting mission and neither plane was carrying the usual bomb load.

"Maybe we'll get lucky or just shake 'em up enough that they'll head for home. Over."

"Hitler's SS Panzer elite... Not *bloody* likely!"

"You with me or not?" Joe called to his wingman with increasing annoyance,

"Aren't I always?" came back a mock contrite response. "I'd follow you to the gates of hell if I have to and today, I think we might just get there. Out!"

Joe ignored the sarcasm and banked sharply, heading for sector 147 about 15 miles north.

"Rowan, this is Red Flight Leader, we're on our way."

Pouring on the coal, the Mustangs raced along at five hundred feet above the hedgerows approaching the raging battle taking place below. Off in the distance, Joe could see a column of smoke rising from a burning Sherman tank and the muzzle flashes of return fire. Pulling up to begin his attack, Joe glanced back, reassuring himself of his wingman's presence - Rob was there.

Leveling off at 3,000 feet, Joe pushed the nose of the fighter down and approached the tank at a steepening angle, his airspeed continuing to climb. Holding tight to the trigger of the control column, Joe watched in

## Robert Brun

frustration as hundreds of rounds of fifty-caliber tracer, armor piercing and incendiary bullets bounced off the exterior armor of the German tank.

Pulling hard to avoid collision with the tree line, Joe swung the Mustang around in a tight turn hoping to have better success with the tank's lighter rear armor, but to no avail. Puffs of smoke and sparks confirmed his aim, but the massive tank continued to move forward unfazed by the attack.

The Tiger's deadly 88mm main gun fired again and up ahead, another piece of allied armor exploded. Joe's frustration grew recalling the directive that had been sent to the pilots from HQ explaining how it was possible to knock out a Tiger by ricocheting bullets off the ground and piercing the vehicle's lightly armored underbelly. At times like this Joe was convinced the advisors at HQ had never served anywhere near the frontline and had certainly never confronted a Tiger in action. The soft earth beneath the tank would only swallowed each round that struck it. There would be no ricochet.

In one final act of determination, Joe maneuvered his plane around 45 degrees to the tank's rear hoping he might, at least disable the vehicle's tread and stop its progress toward the vulnerable Shermans.

Rowan's desperate call continued over the radio, giving a ballgame like play-by-play of the destruction the Panzers were delivering to the 23rd's armor.

"Rob, cover me. I'm going in once more, and this time, at point blank range. Over." Joe barked into his mic.

"Captain, that's crazy!"

The only times Rob ever addressed Joe by his rank were when he questioned his judgment and Rob was certainly questioning it now.

"We can't do anything more with these pea shooters. Our best bet is to radio for reinforcements and hope they can hold out until they get here. Over."

"Rowan's down there and men are dying." and performing a quick snap roll, Joe lowered the fighter to just twenty feet racing across the plowed field kicking up clouds of dirt and chafe with his prop-wash.

Closing in at more than 300 mph, Joe knew he would get only one shot at this so he had to make it good.

Placing the pipper right in the center of the drive cog at the rear of the tank's track, Joe again pulled the trigger releasing a fusillade of bullets that tore apart the tank's exhaust heat-shield and fender, but did little else. Unaffected, the tank kept moving forward, rotating its turret to line up on another target.

Out of the upper right corner of his vision, four bright white flashes streaked passed Joe's canopy jarring memories of Greek mythology. Like the wrath of Zeus' thunderbolts, the fiery apparitions

## My Shadow

raced passed his starboard wingtip with an audible *WHOOSH* converging at the German tank directly in front of him.

Pulling up hard, Joe felt a tremendous jolt and turning his head saw the explosion that nearly struck the tail of his plane. The tank's flaming turret spun end over end to a height of fifty feet before crashing to the ground, a burning mass of twisted metal beside the tank's smoldering chassis. Then something large and dark passed overhead.

Recovering quickly, Joe leveled off at 1,500 feet right along side a grey and green camouflaged British Typhoon fighter. Joe stared in dumb silence then an unfamiliar voice with a distinctive Australian accent broke in over his radio.

"Sorry 'bout that Bruce, but you two Yanks looked like could use a bit of 'elp."

"Uh, yeah..." Joe paused then keyed his mic again. "Thanks."

Still dazed, he looked back to see the second Tiger was also burning and another Typhoon pulling up along side the first.

Larger than the Mustang, the Hawker Typhoon looked like a flying tank. Square and boxy beside Joe's sleek Mustang, the planes had a large gaping maw just behind the propeller that funneled air into its massive 3,000 hp. H-24 Napier-Sabre engine that gave the aircraft a most brutal appearance.

Beneath each wing and below the four long barreled 20mm Hispano cannons were a series of steel rails that stuck out far beyond the wing's leading edge. Attached to these rails were, what looked like oversized bazooka rockets some eight feet in length. Joe had never seen anything so dangerous looking and gave an involuntary shudder.

Back in August, at what had come to be known as the battle of the *"Falaise Pocket,"* on the Cherbourg peninsula of France, RAF Typhoons of the 247th squadron had completely eliminated Field Marshall Von Kluge's Fifth and Seventh Panzer Divisions. In a single afternoon over 200 enemy vehicles had been destroyed.

## Robert Brun

"We over-erd the call for ground support and 'appened to be in the area. 'Is Majesty sends 'is regards. Over."

Flying alongside the two RAF aircraft, Joe continued to stare.

"Yeah... Uh, thank his Majesty for us when you next see him, will ya? Over." All Joe got in reply was a stifled laugh.

A few moments later, the pilot of the Typhoon beside him, turning his head toward Joe, snapped off a sharp British salute and the two RAF fighters peeled away one after the other. Rob then pulled up alongside his flight leader.

"Hot dog! Did you see that?" Joe heard Rob say.

"Yeah, that was really something." Joe replied sheepishly and was not sure if he was feeling relieved or hurt when Captain Rowan broke in.

"Red Flight Leader, this is air controller one. Do you read Joe? Over."

Joe pulled his eyes away from the two rapidly receding British fighters.

"This is Red Flight Leader, Controller one. Go ahead."

"Captain, you got forty-five guys down here all ready to kiss you. Thanks for pullin our fat outta the fire. Over."

Joe paused for a moment watching the two Typhoons vanish from sight. Not wanting to take credit for another man's work, he thought that perhaps this wasn't the best time for an explanation. Without giving it another thought, Joe keyed his mic in reply.

"Copy, Controller one. I'll fill you in on the details over a beer when you get back to the OC. Over."

Listening in, Rob knew he would too.

"Roger, Red Leader, that drink's on me. Over!"

"We'll see, controller one, we'll see. Mad-dog Leader, out!"

# My Shadow

## Chapter 23
## R & R

*Thursday, October 12, 1944: Mixed it up with a few Tigers the other day. Talk about "David and Goliath!" Sure was a good thing those Aussies happened by. Georgia's off to England on leave – R.B.*

Fall arrived in France with the kind of gentle grace Joe hadn't seen since leaving Georgia. There was a crisp feeling to the air and the fields around the air base were filled with the aroma of ripened crops. The markets in the villages nearby were once again brimming with fresh produce and even the mess cooks were experimenting with French cuisine, although not too well.

The AR missions had, by this time, become pretty routine, but no less unpleasant. The German's continued to be pushed back, but resistance was intense, and although the accommodations were comfortable, after three months of this low level fighting, Joe was relieved to find orders to report back to Abington. Colonel Tomlinson had requested a personal progress report, then Joe was on to London for fourteen days *R&R*.

Rob was now on a different leave rotation so this time Joe headed back to England alone.

"Give my regards to the guys back in Abington." Rob called when Joe boarded the supply plane for the flight to Calais and then the short hop across the Channel back to the UK.

"Will do and be sure to hold down the fort while I'm gone."

Joe caught a Dakota from Maupertus Station A-15 and flew back to the 390th BG base at Framlingham. From there, he caught a jeep back to Abington and *home*.

It was an odd feeling being back at the base after so many months away and he felt like a bit of a stranger. Much had changed and the base was full of new faces. After checking in, the next thing Joe did was report to the Colonel.

## Robert Brun

"Sit down Captain." the Colonel ordered seeing Joe enter. "It's good to see you again."

"I'm happy to report the 597th is making real progress in France. I also wanted to personally thank you on behalf of the men for the excellent accommodations provided."

"I thought you might enjoy soaking up a bit of local culture while there. How're things going Captain?"

"AR is a mighty dirty business and far different from the escorts Colonel. At such low levels, aircraft damage is high and we've had our share of wounded, but so far, no casualties. I believe we're getting the job done. Any word on how much longer we'll be pulling AR duty?

"Nothing definite from Command, but my guess is it'll be at least through the end of the year. Now that Paris has been liberated, France has quieted down quite a bit, but Jerry's still putting up a strong fight on his way back home. The Jugs of the 9th are handling most of the AR, but they still need our help. I know the ground support assignment is rough on you and the men, but if we're going to bring this war to an end soon, we need to keep the pressure on." The Colonel looked tired.

"I would have thought" Joe interjected, "the attempt on Hitler's life back in July was a good indicator that the end was near, but it seems the *Führer* has a stronger hold on Germany than anyone knew."

"That or the Nazi High Command is all running scared. I understand the Gestapo rounded up anyone even remotely suspected in the conspiracy. Several thousand lost their lives in that aftermath."

"I've heard. I guess you can't have the rational people hanging around mucking things up with sanity." Joe added with sarcasm.

There was a pause and neither man spoke.

"You heard about the debacle in Holland, I take it?" the Colonel asked referring to *Operation Market Garden*. "Off the record, it makes me sick to think of all those men wasted, and for what?"

"I only know what I read in *Stars and Stripes* Sir. A calculated risk I suppose. A real shame though, just when it was looking up. Things don't always turnout as planned Colonel, you know that."

"Calculated cock-up" is more like it." Tomlinson spat, "Intelligences reports were clear about the enemy build up in Holland and that information was ignored. I think some of those at HQ smell victory and they're getting sloppy."

"Can you really blame them Colonel, we're all getting tired."

"Tired..." The Colonel's voice trailed off.

"There is some good news though. Reports are coming in that the city of Aachen was taken two days ago, the first German city to fall. Troops entering the outskirts reported hundreds of refugee Germans fleeing the town even as the fighting was still going on around them. It

## My Shadow

looks as though their war has finally come home. The Nazis may continue to fight, but it appears the German people are starting to wake up to reality."

"I hope you're right, Colonel, I'd sure like to go home."

"So would we all." There was another pause. "But enough shop talk, you've got two weeks of well deserved R&R coming."

"Thank you Sir. I'll leave my full report with the Adjutant. As for the men, I guess the Forward AR can tough it out a while longer and as for me I've got a train to catch. Colonel!" Joe stood and saluted.

"Captain," the Colonel said and picking up a pen and paper, began to write. "Here is the name of a friend of mine. He and his wife have a small place on the outskirts of London. It's beautiful this time of year and you might want to consider it for part of your leave. Just mention my name and they'll treat you like the King." Tomlinson handed Joe the address.

"Thank you Colonel, but I wouldn't want to intrude."

"I assure you, it's no intrusion. Gordon Kingsley and I go back to my days with the *Eagle Squadron*. You might say he `owes me' and I'm sure he'd appreciate a chance to even things up."

"Thank you sir, I might just do that." Joe stuffed the paper into his shirt pocket.

Joe caught the lorry to the train and once on board, slept most of the way to London. Arriving at the station, he grabbed his kit and pulling the address for Diana's brownstone from his jacket, flagged a cab and handed the address to the driver. He'd decided to pay them a visit while in London and was looking forward to catching up.

Peering out the window while the driver fought the afternoon traffic through the heart of London, they drove pass the *Keg and Flask* where Joe and Reg had enjoyed their meal that first day. A strange anticipation, the likes of which Joe had never before felt began as the cab neared the street where Diana lived.

Pulling around the corner, Joe saw the street was cordoned off with barricades, he rechecked the address. Instructing the driver to wait, Joe got out of the cab and questioned the *Bobbie* guarding the area.

"What happened?" Joe asked the officer standing behind the yellow and black barriers blocking the road. Looking passed the guard Joe could see rubble strewn across the street and fallen trees laying along the sidewalk. The air was heavy with the smell of wood smoke and charred earth.

"*V-2 landed ere last night,*" the officer answered. "*Bloody awful tings day are. Never ear a ting till after day `it! Take out a `ole city block, day do.*"

# Robert Brun

"Vee-two?" Joe asked puzzled, his anxiety increasing. "What's a vee-two?"

*"New Jerry `Super Weapon' I'm told. Guided rockets. Fast as ell, dat's all I know.*

*"No...* Diana?"

In a sudden panic, Joe pushed passed the police officer.

*"Ay now, `old on dare."*

Taking off down the street, Joe leaped over blocks of broken masonry until he came to the place where Diana's brownstone should have been only to find a pile of rubble. Directly across the street was a deep crater twenty feet deep and still smoldering.

The *Bobbie* ran up beside him and Joe, grabbing him by the shoulders, shouted.

"Diana... Diana and Harry Griffin. Where are they?" Joe was yelling now. "THEY LIVE HERE. WHERE ARE THEY?"

*"I don't know. I've only been ere since dis morning. Didn't ere anything about survivors."*

Joe went silent, staring back at what remained of the building. Then he got control of himself.

"I'm sorry." He said releasing the officer and seeing only bewilderment on the man's face. "She's a friend of mine."

Joe walked back to the barricade and collecting his duffle, climbed back into the waiting cab. Pulling the slip of paper Col. Tomlinson had given him, he handed it to the driver without a word.

Gordon and Heather Kingsley turned out to be a most delightful and understanding couple. In their mid-thirties, they put any feelings Joe had about intruding to rest and welcomed him like family. Their rural estate outside of London proper in East Sussex sat next to a small meadow alongside a brook. Joe's recent experience in London had emboldened him to seek out the unfamiliar couple on the Colonel's recommendation.

## My Shadow

A small room above a cottage that doubled as the garage served as Joe's accommodations for the stay. The apartment included a small *cooker,* private bath and well-stocked larder. The Kingsley's made it clear that he was under no obligation to socialize and could come and go as he pleased.

For the first few days, Joe did little more than sleep, catching up on rest and enjoying his solitary afternoon tea. A small porch off the back of the cottage overlooked the brook and was a perfect oasis, but everything seemed to remind him of Diana and Harry. She was in his thoughts and even infiltrated his sleep.

In the dreams he would be with Diana, then she would transform into Mo, with Harry coming and going as he pleased, always looking at Joe with his big innocent eyes. And each dream ended the same way, with a sudden explosion and a scream that jolted him awake in a sweat.

By the beginning of the second week, the dreams were becoming less frequent and Joe started leaving his room to explore the area around the cottage. One afternoon, while admiring the sports car in the garage below the apartment, Joe was greeted by Gordon.

"It's a 1939 MG TA Midget." Gordon announced, a certain pride in his voice. "Are you an admirer of fine automobiles?" He asked seeing Joe looking at the car.

"Where I'm from, the fanciest vehicles all pull plows. She's a real beauty though."

"I bought it just before the war to celebrate *"Peace for our time."* but haven't driven it much, what with the shortage of petrol. Would you care to take her for a spin?" Gordon offered.

"Honestly?"

"Sure, I've been storing a bit of my petrol ration for over three years now, saving up for a special occasion and I think now would be as good a time as any. You just wait here."

Joe watched Gordon disappear out of the garage and return a few moments later with a gas can in each hand.

"I'll fill the tank while you take it down off the jacks."

Twenty minutes later, and with Joe behind the wheel on Gordon's insistence, the two men raced out the drive and onto the country road heading toward Dover. Driving on the wrong side of the road and shifting with his left hand took a bit of getting use to, but by the time they'd reached the outskirts of Ticehurts, Joe was driving like a regular *Barney Oldfield.*

Gordon directed Joe along the winding roads of South East England until they pulled the car off to the side atop Beachy Head and the famous White Cliffs of Dover.

## Robert Brun

There, the two got out of the car and walked to the edge of the Cliffs looking out across the English Channel. Joe had flown over this stretch of water many times, but had never before seen it from the ground. The autumn afternoon was crisp and clear and Joe could see Calais, France from where they stood atop the chalk white cliffs.

"For over a thousand years this stretch of water has protected Britain from invaders, but now that's all changed. Never again will England feel as safe or secure as before 1939." Gordon said with a sigh. "I'm afraid the Empire I grew up under is gone forever."

"They'll always be an England!" Joe quoted the popular song by Vera Lynn.

"Yes, but what will that mean after all this is over?

Joe didn't reply.

A moment later and without a further word, Joe tossed the keys to Gordon and climbed into the passenger's seat for the return trip.

Joe rode in silence for most of the way back, watching the countryside slide by and listening the flawless way Gordon operated the gearshift. It was clear he loved this car.

"How do you know the Colonel?" Joe asked hesitantly.

"G.W.?" Gordon asked with a chuckle then went on. "He was one of the first of you Yanks to come over in 1940. Part of the 71st 'Eagle Squadron.' By February 1941 when they became operational, I was stationed at Ipswich and he was next door at Martlesham Heath in Suffolk.

We were all pretty skeptical of you Yanks when you first arrived, not being proper British and all, but once we saw what you could do, we changed our tune right quick. Besides, we sure needed the help."

Gordon downshifted the MG and took a hairpin turn without batting an eye.

"I met G.W. in a pub outside Croydon after one of his first missions. We had scrambled earlier that morning to intercept a flight of Junkers over Hornchurch and when we got there, G.W. and the rest of the 'Eagles' were already mixing it up with Jerry. With the cry of '*Tally Ho'* from our squadron leader, we dived into the fray.

"I mixed it up with a few of the Hun fighters before breaking into the bombers and am embarrassed to admit I got a bit fixated on a Ju 88 when I felt the thumping of bullets against my seat back armor. Ol Jerry was giving me a right good pasting. I was sure I was a goner, but before I could take evasive action, I looked in the mirror to see the 109 explode! Seconds later a *Spitter* flew by sporting GW*T and bearing the crest of the Eagle Squadron."

"Back at the base, I vowed to find the chap who'd save my life and buy him a drink. I told G.W. that first night; if there was ever anything I could do for him, to just ask. Since then, we've each gone our separate

## My Shadow

ways. He's a Colonel in the USAAF and I've moved into British Intelligence, but we manage to keep in touch."

Gordon pulled the MG into the driveway and shut off the engine.

"I don't mean to intrude old man, but the Misses and I were wondering if you care to dine with us this evening? Nothing fancy mind you, but I thought you might appreciate a change of scenery."

That evening, Joe joined Gordon and Heather as their guest. After a wonderful dinner of mutton, Gordon and Joe retired to the drawing room for cigars and brandy. After some light conversation Joe asked.

"How would one find out about *blitz* casualties?"

"What do you mean old boy? Is there someone in particular?"

"A woman and her son, Diana and Harry Griffin, I stayed with them on my last leave back in June. Their house was hit by a V-2 the night before I arrived."

Gordon thought for a minute then said.

"I have a friend at the records office. Let me make a few calls in the morning and see what I can come up with." He wrote down the names.

Feeling comfortably full and encouraged, Joe and Gordon talked until neither man could keep his eyes open. Joe then excused himself, thanked his host and retired to the cottage where he slept until ten o'clock the following day.

That morning while enjoying tea on the small porch off the back of the cottage, Joe heard a knock at the door and opened it to see Gordon standing there. He could tell by the expression on his face that it was not good news.

"I'm sorry," Gordon began softly noting the anxious look on Joe's face. "Diana and Harry Griffin both perished in the V-2 attack."

Joe stiffened slightly.

"Thank you Gordon." Joe said and closed the door.

Early the next morning Captain Joseph Dyer left the Kingsley's in East Sussex and headed back to France.

# Robert Brun

## Chapter 24
## Thanksgiving

*Thursday, November 23, 1944: Georgia's back from R&R and something's up. I can't quite put my finger on it, but it must have happened while on his leave. He seems distant and confused. – R.B.*

    The fall weather turned as confused as Joe's feelings for Maureen. One day, rain and fog, the next day bright sunshine followed by windy and cold. Joe had noticed since his return from London, he had started think of Mo by her full name, no longer sure their bond was as strong as it had once been. He had no idea what was happening back in the States and no way too know. With the distances involved and the slow, irregular mail delivery it was weeks before he could hope to get a reply to any letter he might now send. The occasional letter from her, when one did come, remained platonic, retelling events from back home and saying nothing of how she felt about him. Of course there was always some mention of *Doug!* *Doug* this and *Doug* that, he wished they could just sit and talk like they used to, but that was impossible for the time being. He would simply have to wait.
    Joe was now certain about why he had gone to London, although he hadn't known it at the time. He had gone looking for Diana, trying to recapture the sense of peace and belonging he'd felt back in June before the invasion. But what had he really hoped to find, and what would he have done with it if he had? He couldn't say. Now none of this mattered; Diana and Harry were dead and with them had died any fantasy he may have had.
    These conflicted feelings were all new to Joe, and well beyond the experience and training of a twenty-one year old fighter pilot from Georgia. As a Squadron Commander, he was a leader of men, but was he

## My Shadow

in control of his own feelings? He longed to be certain of something again, to feel he belonged to something and now, more than ever, he wanted to hear from Maureen and know how she felt.

The reelection of President Roosevelt for an unprecedented forth term earlier in the month had insured political consistency on the home front, though Joe didn't know much about the new Vice President, Harry S. Truman. A major allied offensive was underway since the 1st that pushed the *Wehrmacht* back to the Roer River and had been keeping the 597th forward AR busy all throughout November. Now, even that seemed to have ground to a halt. The remaining German troops had dug in to defend their homeland. For Joe, everything felt up in the air both here and back at home.

Joe stood at the head of the long table in the dining room at the château. As the senior officer present, it was his job play the role of patriarch and carve the Thanksgiving goose provided, in substitution for the unavailable turkey. The mess cooks had done their best to provide a traditional Thanksgiving meal for the pilots and crews, but Joe's thoughts remained four thousand miles away.

The château's large dining room was filled with tables and chairs to accommodate as many of the base's personnel as possible. All of the pilots and as many of the ground crews as would fit had been invited to share the meal and give thanks on this most American holiday. Only those on immediate duty were absent.

Despite all these festivities, Joe was having a hard time finding something to feel thankful for. Nothing was the way he wanted it to be. He was 4,500 miles from home and the people he loved, facing death and destruction almost daily, unsure where he stood with Maureen and now two more people he had felt close to were dead.

Slicing the goose while the men passed plates of food back and forth, Joe heard a rapid ringing sound and looked up to see Rob standing at his seat tapping his wine glass with a fork. The room fell silent.

"Captain Dyer, if you'll excuse me for a moment. Before we begin this magnificent feast so generously provided by Uncle Sam and assembled by the culinary geniuses of the 597th's kitchen staff." Everyone looked over at mess crew, standing by the door to the kitchen wearing makeshift chefs hats constructed of tag-board.

"I believe Lieutenant Taylor has something he would like to say. Lieutenant?" Rob yielded the floor to a surprised and reluctant Lieutenant Taylor who, being put on the spot, stood slowly and cleared his throat.

"Ahem... Gentleman... and I use the word loosely." There were chuckles among the men assembled. "Today is Thanksgiving, a day of reflection to give thanks for that which we have received over the last year.

## Robert Brun

"Like many of you, I too am having a hard time thinking of anything to be thankful for." and there was a murmur of agreement from the men assembled.

"For the last year and certainly for the last four months since our arrival in France, we have been facing a most unpleasant task almost daily. Flying into harms way, risking our lives and getting shot at by an enemy that doesn't seem to have the sense to know when to quit. Our lives here are far from ideal and if you're anything like me, you're a long way from where you want to be and the ones you love."

"Yeah, and we've all seen *her* picture painted on the nose of your plane. Hubba-hubba!" Wilson shouted followed by wolf-whistles from the rest of the men.

"Okay... okay, enough about that!" and looking only slightly embarrassed, Taylor continued. "What I really want to say is, that when I stop to think about it, what we're doing here may stink, but it is important, and the fact is we're still here and God knows anyone of us could be gone tomorrow." There was a pause and no one spoke. Then, with renewed vigor, Taylor went on.

"But we are here today. Alive, together, with a roof over our heads, sharing this meal and for that, if nothing else, I for one, am truly thankful. Thank you." Blushing slightly, Taylor retook his seat.

The room was silent while the men contemplated what they had just heard. Joe considered the words carefully. Taylor was right, for the moment, this was all any of them had and no one could say for how long even this would last. He, like all the men there, had volunteered for this and they had to see it through to the end no matter what the cost.

"Eat, drink and be merry, for tomorrow we die." Joe thought.

Joe looked around the room at the faces of the men he had been serving with since his arrival, and there weren't many left. Those, along with the replacements newly arrived, for the moment, were sharing this brief time together. Glancing over at Rob who absent-mindedly ran his finger around the rim of his wine glass making it ring, Joe smiled. Then Chaplain McNulty stood and asked the men to bow their heads for grace. Looking down, his hands folded in his lap Joe too, like Taylor and the other men, had now found something to be thankful for.

# My Shadow

## Chapter 25
## The Bulge

*Saturday, December 16, 1944: Christmastime again and things have quieted down. Weather's been awful, but at least we're getting a break from the Jerrys. Maybe the Brat's are finally throwing in the towel? - R.B.*

The small tree looked ridiculous standing on the table over by the château's window, but still it made Joe smile. Raisins strung on thread and small pieces of aircraft aluminum adorned the boughs. Someone over at the machine shop had hammered a star out of a scrap piece of fuselage, and the spent shell casing added an interesting touch. Though already Joe's second Christmas since shipping out, this time, it almost felt like home.

It was a quiet week, and the weather had been awful. There hadn't been much flying, but then there wasn't much need too. The *Wehrmacht* was in full retreat. German occupied towns were falling like dominos before the advancing allied armies and rumors abounded that the war would be over by Christmas. Joe sure hoped they were right, although he couldn't share the same enthusiasm. None of that mattered now - this was Christmastime, a time for joy and celebration.

That evening Capt. Porter, leader of Blue Flight, had taken over the battered upright piano in the château's parlor and was leading those assembled in a chorus of Christmas classics. *"The First Noel, Away in a Manger"* and *"Ding Dong Merrily on High"* flowed out of the men with a genuine warmth that reminded Joe of Christmases back in Georgia. Even the cooks were making plans for a veritable feast for the 25th.

Half way through *"Good king Wenceslas"*, the front door flew open and Rob came rushing in all excited.

## Robert Brun

"Hey everybody, it's snowing!" he announced and launching a snowball across the room, hit Capt. Porter right in the side of the head.

"Oops!" Rob blurted with a stunned expression, obviously not having intended to actually hit anyone.

Porter slammed the piano lid shut and, knocking over the stool, took off after Browning who'd beat a hasty retreat back out the front door.

Never wanting to miss a good fight, everyone in the room followed the chase and watched while Porter, like a line backer preventing a touchdown, caught up with Browning tackling him onto the snow-covered field. Unable to just stand by and watch, the rest of the men joined in, knocking each other down and throwing snowballs back and forth. Two teams soon formed and a fierce snowball fight commenced.

Ducking down behind a vehicle parked nearby, Joe picked up handfuls of snow and carefully packed them into two round balls. Sneaking slowly around a deuce-and-a-half, Joe saw Rob creeping up on Capt. Rowan hiding next to the maintenance shed. Raising his arm to throw, Joe let fly with first one, then the second projectile catching Rob between the shoulders and again in the stomach when he turned. Rob spun around, melodramatically grasping his gut and falling to the ground.

"Nice...very nice." Joe applauded approaching Rob's twitching body. "I think you've missed your calling. I'd say *can* the art stuff and move to Hollywood."

"What!?" Rob retorted indignantly, "Bite your tongue."

Browning accepted Joe's hand-up just as a fusillade of snowballs rained in from all directions. Collecting themselves together, Joe and his wingman retaliated, firing back volley after volley of snowballs in rapid succession.

Seeing Capt. Porter over by the mess hut, Rob let one fly only to see Porter smoothly sidestep the throw, which proceeded to knock Colonel Tomlinson's cap off his head. Rob froze surprised to see his Abington CO, while a startled Porter called out, "A ten-Hut!"

Instantly, the battle came to an abrupt end while the Colonel slowly bent down to retrieve his cap. Why was the Colonel here anyway? They hadn't seen Tomlinson in over five months.

"All you men," Tomlinson's voice boomed. "I want everyone in the briefing room on the double!"

Not five minutes later every pilot of the 597th Forward A.R. Division, as they were now classified, was standing at attention in the briefing room.

Colonel Tomlinson strode to the front of the room in his usual fashion, his uniform cap dented and wet on one side. Immediately taking charge, he cleared his throat and began.

## My Shadow

"Men, we just received word that three German army units have made an attack along a fifty mile front through the Ardennes forest outside of Bastogne. It's been reported that 21 divisions, that's 250,000 troops, along with eight Panzer divisions are involved." There was muffled discussion throughout the room.

"Intelligence estimates more than 500 tanks and artillery pieces have been moved into the area. Details are still sketchy, but it would appear that this is a major counter offensive with several units of German SS Panzer elite. The 8th, 9th and 15th air forces have all been alerted, but because of the weather everything is grounded and HQ says it looks like it's going to stay this way for the foreseeable future. We are to stay on full alert, however, and counter attack the instant the weather clears. In the meantime, I'll be assuming command and keep you posted so stay ready. Dismissed!"

"The Ardennes Forest?" Rob asked approaching Joe from behind, "Isn't that the same route the *Brats* already used before... twice?"

"I think you're right, but until this weather clears up, flying will be close to impossible."

"We may soon find we'll be asked to do the impossible." Rob replied, looking worried.

Joe looked out into the falling snow with a new feeling of foreboding. This was not the kind of "White Christmas" he had been hoping for.

Because of the dense forest of the Ardennes, the terrain in the area was thought to be impenetrable by a large army, so the salient had been lightly defended with green troops. Since the German break through, hundreds of these troops had already been killed or captured and the air force could do nothing at all to help! Despite the season, that night, no one felt like celebrating while the men waited for the weather to break.

The following morning conditions weren't any better and Joe was surprised when orders came through at the briefing to prepare Red Flight for an A.R. mission. Approaching his plane now weighed down by the two 500lb delayed fuse bombs, he could hardly make Zeke out through the mist. It was clear that weather conditions were far from ideal, but he and Red Flight were under orders to fly anyway. Zeke was quiet that morning and seemed to bite down even harder on his cigar stump. Before closing the cockpit, he looked at Joe.

"Be careful today will ya, Cap?"

"Just like always Zeke."

It seemed to ease his mechanic somewhat and removing his cigar, Zeke spat on the ground.

## Robert Brun

It was a hard climb through the clouds and visibility was zero right up until the four planes broke out atop the heavy overcast. Red Flight formed up heading toward the assigned coordinates while Joe looked for a hole in the clouds. Somewhere down below the endless expanse of white he knew the Panzers lay, but where and how to get to them was the problem. Flying along through the assigned sector, Joe at last spotted a small opening in the cloud cover and radioed the rest of Red Flight.

"Red flight leader to Red Flight. Browning, you, Wilson and Taylor stay up here, I'm going down to have a look around, but stay on this frequency. If I give a shout, I want you three down with me PDQ!"

Joe scanned the sector map reassuring himself he wasn't about to fly into a hillside and dropped *MoJo II* down into the clouds below.

The world outside Joe's canopy turned a solid white and he forced the idea of unseen obstacles from his mind, but soon the mist began to darken and the vague shapes of trees against a snow covered landscape started to appear.

Getting his bearings, he saw there was about 200 feet of airspace between the treetops and the overcast. Slowing his fighter, Joe followed the tree lines looking for anything moving when a line of tracers appeared from the left. Banking hard, Joe turned the ninety-degrees right into a line of tanks larger than he had thought possible.

"Mad-dog Red leader to Red Flight, I've got multiple armor and could sure use some help. Over."

"Mad-dog leader, we're on our way."

Joe didn't wait for the others to arrive, but brought the Mustang around, lining up for a low pass and trying to ignore the ground fire that got closer by the second. Paralleling the roadway, he released the two 500 lb. bombs and ten seconds later heard the *whump* of their explosions as he flew over the endless enemy column. Turning 180 degrees to repeat another pass, the remainder of Red Flight came barreling down out of the clouds releasing their bombs over the column of tanks. The explosions burst skyward and tanks were exploding, some flipping over onto their backs. Joe made a final pass over the column firing his guns then ordered Red Flight back into the overcast and home.

After touchdown back at the base, inspection showed that Lt. Taylor had taken a 20mm round in the left wing and Wilson's plane was showing some bomb damage to the tail, but to his relief, none of the pilots had been hit.

The following day, the Mustangs of the 597th made another attempt to engage the enemy, but this time the fog was just too much for them and after about half an hour they were forced to return to base. As a precaution, the ground crews had lit flares along the edges of the runway, without which none of the planes would have made it back.

## My Shadow

For the next four days nothing flew out of the 597th and the weather went from bad to worse. Snow was replaced by dense fog and everything continued to be grounded. The troops defending the Ardenne would have to fend for themselves.

By this time, Generals Patton, Bradley and Montgomery were all trying to bring up reinforcements from 100 miles away, but with the roads frozen and the daytime temperature never exceeding 20 degrees Fahrenheit, it was slow going.

Finally on December 23rd, an anti-cyclone brought with it a high-pressure system that cleared the skies. Red Flight, along with the rest of the 8th and 9th Air Forces based in France, were bombed up and at first light, raced across France toward Belgium and the *Bulge* with only one goal, to break the offensive and push the Germans back.

German armor stretched out for miles and anti-aircraft fire was at times heavy, but without German air support, the Panzers had no chance against the American *Jabos*.

Everywhere Joe looked there were Thunderbolts and Mustangs, all carrying bombs. The devastation was swift and brutal and the *Wehrmacht* was now paying the price for their break through. The tide of the battle had turned.

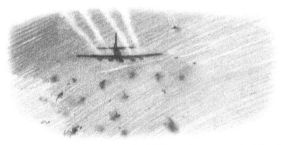

That Christmas Eve, over 2,000 bombers escorted by 900 fighters flying from England bombed the airfields around Frankfurt completely disrupting German communications throughout the entire area of the *Bulge* and putting an end to what was the last major German offensive.

Joe landed back at their forward base at 16:30, in the fading light on Christmas Eve. A light snow was falling and his prop-wash scattered the flakes when *Mojo II* spun around and onto the hardstand. Climbing up on the wing, Zeke opened the canopy and to Joe's pleasure and surprise, handed the Captain a steaming hot canteen of whiskey-laced eggnog.

"Merry Christmas, Captain." Zeke said.

"And Merry Christmas to you, Sergeant."

Zeke smiled and spat on the ground.

# Robert Brun

## Chapter 26
## Base Attack

*Monday, January 1, 1945: Some break! All that quiet turned into a massive Brat sneak attack! Finally pushed them back though when the weather cleared. Man, this job gets tougher everyday - R.B.*

    The last few days of clear weather gave way to a heavy snow late in the afternoon of the 31th, shutting down all airborne operations. Joe returned from the mission of the previous day exhausted and aching all over. Skipping dinner, he headed to his room, fell asleep and awoke the next morning feeling miserable. Having no appetite, he passed on breakfast and checked in at sick call while making his way to the base medical hut. The head cold he had been ignoring since Christmas had finally become a full-blown case of the flu.
    Waiting in Major Lewis' sickbay, Joe sat on the edge of the examination table shivering. He had been suffering a sore throat for the past week and a few fits of sneezing, but so had most of the other pilots in the squadron. Now, his body would no longer allow him to ignore the symptoms.
    "You've been pushing yourself pretty hard for a while now, Captain," Major Lewis said shaking the thermometer while Joe pulled his uniform shirt back on, "And don't bother complaining, you need a few days rest and, that's exactly what I'm prescribing," handing Joe a bottle of aspirin. "So save your belly aching for the Germans and get to bed."
    Joe started to say something in his defense, but he was just too tired to follow through.
    "That's an order, Captain!"
    Bundled up in his topcoat, flight jacket, watch cap pulled down over his ears and collar turned up against the wind, Joe made his way back to the château. The cold weather of the last month along with the almost

## My Shadow

non-stop flying to the *Bulge* had caught up with him. He could feel the aches deep down in his bones. Even his hair hurt.

Some of the mess crew were beginning to decorate the dining room and Joe was reminded that today was December 31st. Tonight was New Years' Eve. He made his way up the stairs to his room and shut the door.

Pulling off his uniform, Joe dry swallowed four aspirin tablets and tossed some more coal onto the fire then wrapping his shivering body in blankets, he lay down on the bed and despite the activity one floor below, Joe was soon out cold and slept all that day and through the night.

The Messerschmitts and Focke-Wulfs raced along at less than one hundred feet above the snow-covered fields of Hunsrück while the eastern horizon continued to brighten with the dawning of the New Year. The remainder of the *Gruppe* flew in tight formation skirting the French border, staying below the hillsides and beneath the American's early-warning radar. JG-48 was only one of more than twenty-two *Jagdgeschwader* flying that morning. One squadron of the over 875 fighters sent out by the *Luftwaffe*.

Following the Saar River to where it met the Mosel then turning east into France, the *Oberstleutnant* leading his flight hoped to catch the American forward airfields off guard and recovering from their revelry of the night before. Operation *Bodenplatte,* or Base plate, was a New Year's gift from the *Führer* designed to rid the continent of the American and British *Jabos* for good.

Approaching the French château near Metz, an airfield that had once been their home, the fighters of JG-48 released their external fuel tanks and spread out across the horizon ready to deliver the *coup de gras* to the unsuspecting pilots of the 597th. The *Oberstleutnant* was in a foul mood having already lost three of his *Gruppe* to his own flak. In an effort to keep the mission a secret, German aerial defenses had not been alerted. Now he would exact his revenge.

Joe had no idea how long he had been asleep when he awoke to the sound of shouting and people running around downstairs. Confused, and still only half conscious, he thought he heard the sound of aircraft engines whining as if in a dive followed by the unmistakable chatter of machine gun fire. Then his room went crazy!

Not certain he wasn't in the midst of some surreal fever dream, Joe lay there watching objects throughout the room explode, shattering before his eyes. It only took a second longer for him to realized what was happening; his room was being strafed, but how and by whom?

# Robert Brun

Rolling out of bed, Joe pressed his body flat against the floor while chunks of plaster fell from the walls and ceiling. When it appeared the shooting had stopped, he got up and ran to the shattered window to see a low flying plane pass close enough to cast its shadow where he stood. What the hell was going on? He wasted no time pulling on his trousers, boots and topcoat and made his way through the broken glass, running downstairs and out the front door into the light of the New Year.

What he saw it was total pandemonium. Pilots and ground crews were running in every direction and several hangers were on fire. Out on the airfield, two Mustangs were burning and above it all, were the unmistakable silhouettes of German fighters diving and climbing from every direction.

Overwhelmed by the chaos before him, Joe wandered out onto the apron of the airfield, barely getting ten feet before being tackled by someone he hadn't seen and landing hard in one of the slit trenches that bordered the runway.

"What the hell do you think you're doing?" he hollered at the soldier who now lay on top of him pressing his face into the frozen earth of the trench.

"Happy New Year Georgia, now shut up and keep your head down before you loose it. We're under attack and I've got no plans to get my ass shot off here on the ground." Joe heard Rob shouting to him above the din.

Among the staccato reports of the base's anti-aircraft fire were the repeated *whumps* of exploding bombs and raining debris. The two men covered their heads with their hands not daring to move.

It was hard to gauge just how long the attack had lasted, but the bombing around them did eventually stop. The enemy fighters were gone and Joe and Rob cautiously peered out from the trench where they lay.

## My Shadow

Smoke was rising from several of the hangers and the château. The runway was pitted with craters, dirt and debris everywhere.

Walking across the base, Joe watched while members of ground crews and mechanics began organizing work details to put out fires and salvage scattered equipment.

Ignoring his fever, Joe began helping some of those who had sustained injuries during the attack, pulling men out from under collapsed buildings and overturned vehicles. Over by the maintenance hanger, Joe was relieved to see Zeke barking orders and swearing in his usual fashion, rallying the mechanics as he did so. Not wanting to seem too obvious, he made his way to the revetment where *MoJo II* was kept and with the exception of a coating of dirt, she seemed to be intact. It was then that Joe saw *Shadow* engulfed in flames and burning violently, Rob silhouetted against the conflagration. Ignoring the intense heat of the burning fighter, Joe walked over to where Rob stood and as he approached, heard Rob utter a single word.

*"Paula!"*

Rob showed no reaction when Joe placed a consoling hand on his wingman's shoulder and Joe knew there was nothing to be said. He left Rob to mourn the loss in his own way.

Once again shivering with fever, Joe headed back to his room and despite the bullet holes and shattered panes of glass he wrapped himself in the blankets and soon fell back into an uneasy sleep.

For the first two days of 1945, Joe lay in his bed alternately sweating and shivering while his fever rose and fell. In and out of consciousness he experienced almost continuous nightmares while his body fought off the virus. Hallucinations that he was trapped in his plane, his hands and face on fire plagued his sleep and he heard himself scream feeling the burning flesh. In his unconscious state he thrashed about, the plane spinning in his mind, waves of dizziness and nausea alternately washing over him. Visions of Mo, Diana, Harry and Cushman taunted him constantly during his illness.

Finally, on the morning of the third day just before sunrise, the fever broke and Joe awoke with a start followed by a wonderful sense of relief. His body no longer ached and the damp bedding now felt cool against his skin. After the struggle of the last two days, he was completely exhausted and rolling over, once again fell into a deep dreamless sleep.

The following morning, weak and still disoriented, Joe made his way downstairs to the dining room, not sure which events of the last days had really happened and which he'd only dreamt. The shattered remains of the château's kitchen confirmed that not all had been in his head.

Exiting the building, Joe followed the path over to the maintenance area where the blackened shell of what had once been his wingman's plane lay in a charred heap.

"A sad end to a gallant steed." Joe thought and wondered how Rob was doing. Only the knowledge that Rob had not been in it brought some relief to the sight of the twisted wreck. The rest of the airfield was littered with burned out fighters, some still smoldering.

Back inside the château, Joe entered the dining room and was greeted by several of the other pilots who inquired about his condition.

"How you feeling, Captain?" asked one of the cooks, who passed Joe a cup of steaming coffee.

"About the same as the base, I'd say. I think we both got the shit kicked out of us."

"You can say that again." the cook replied. "The kitchen took a direct hit. We've been cooking out of the back of a deuce-an-a half for the last few days until we can rebuild."

"So we can expect an improvement in the chow till then?" Joe asked the Corporal grabbing a tray and heading toward the chow line.

"Yeah, you're feeling better, Captain," remarked the cook.

Halfway down the line, Joe noticed a Private ladling out oatmeal from a pot with a large *swastika* painted on the side.

"What's this Private?" Joe asked pointing at the stenciled marking with his spoon.

"Well Captain..." the Private started modestly, but before he could finish, Rob's voice chimed in from behind him.

"That, my dear Georgia, is the first kill ever made by a cook in the history of the 8th Air Force. Cookie here, while the rest of us were running for cover, grabbed one of the base's fifties and shot down a 109."

Joe looked over at the cook in time to see him blush red.

"Good work Private." Joe said to the boy.

"Thank you Sirs." and the Private beamed with pride.

Having eaten little over the last few days, Joe was famished. He filled his tray with powdered eggs, bacon, toast and even the SPAM looked appetizing. Sitting down at the nearest table, Rob followed and sat down across from him with his cup of coffee.

"Better take it easy with that chow Georgia or you'll end up back in the infirmary." Rob warned.

"I'm so hungry, I can even eat this stuff." Joe remarked as he began devouring his plate of food. "By the way, what have I missed the passed few days?"

## My Shadow

Rob was suddenly quiet and lowered his eyes to study his coffee cup. There was a long pause then Rob let out a sigh.

"Harrison's dead!" Rob announced forcing out the words.

Joe put down his fork and stared over at Rob.

"No! He just got married on his last leave. When? How did it happen?"

"The day before yesterday, during takeoff. He was bringing up the rear of green flight and caught the edge of a bomb crater that was missed by the repair crews. The right main wheel collapsed and the Mustang flipped. The crash must have ruptured the fuel tank because the whole plane went up in flames before the fire trucks could get to him."

"Has his wife been notified yet?"

"The CO's sent the usual letter... *We regret to inform you, blah, blah, blah!*"

Joe pushed his plate away. He'd suddenly lost his appetite.

After breakfast, Joe showered and shaved and put on a clean uniform. Feeling almost human again, though still weak, he was nevertheless anxious to get back on duty and into the air.

Strolling over to the runway Joe noticed several brand new Mustangs sitting on the perimeter track. Replacement fighters were already being flown in from England and repairs were continuing on the hangers and runways. The attack had been violent, but damage was light.

A little farther along, Joe was curious to see Rob's crew chief Tom and several of the maintenance crew standing around looking up into the sky. Noticing the sound of a Merlin engine at full emergency power, Joe followed their gaze and saw a single P-51D performing acrobatic maneuvers high overhead.

Rolling over into a dive, the Mustang flew straight down, finally pulling out a mere fifty feet above the airfield. The plane then executed a four-point roll over the runway and zoom climbed straight up. Gradually coming to a stop, it fell over onto its back and into an inverted spin.

Joe watched the fighter slowly nose down, recover and disappear behind the stand of trees at the end of the field. A few minutes later the Mustang, flaps and wheels down, flared out over the runway and made a prefect three point landing.

Despite the cold temperature, the sun felt good against Joe's face and he stood there in relative comfort watching the shiny new Mustang taxi over and come to a stop where the men stood. The bubble canopy slid back and Rob peered out from the cockpit looking like the cat that had just eaten the canary.

"So, how'd you like it?" Joe asked nonchalantly.

## Robert Brun

"She's real *Jive*." Rob shouted back and began telling his flight leader about all the characteristics of the P-51D that Joe had been flying for more than six months.

"So it only took a *Luftwaffe* bomb to finally pry you out of the *Shadow*, huh?" Joe said playfully admonishing his wingman. "By the way, I didn't know you could fly like that!"

"The *Shadow* knows." and tossing Joe his flight helmet, Rob turned and began discussing the cowling art with his crew chief.

Later that same afternoon, Joe entered the château for dinner and was met by an unusual sight. Two armed British Corporals stood on either side of a young man dressed in a *Luftwaffe* flying uniform. The pilot, covered in soot, black oil and mud, sat nervously eating while his eyes darted around the room. Judging by the way he ate, it looked as though he hadn't tasted food in a very long time and sat devouring his tray of food.

An eerie silence fell over the room when the German pilot stood. Everyone present watched as he turned and keeping his eyes straight ahead, was escorted out of the château and into the back of a waiting British lorry.

"They picked him up this morning hiding out behind the latrine." Rob said, filling in the details. "He crash landed out past the vineyard property and has been on the lamb for the last four days. I guess good sense finally got the better of him, realizing there was no way to make it back to his own lines. That and hunger."

"Did you see his face?" Joe asked Rob who sat down at the table "I've never seen such fear in a man's eyes before."

"Yeah, I noticed that too. I hear the Germans tell their pilots that we *eat* our prisoners. I guess it's Göring's way of keeping them fighting till the very end.

# My Shadow

## Chapter 27
### G-suits

*Friday, February 2, 1945: Shadow's a wreck! Dammed Brats hit the base New Year's Day and busted everything up pretty bad. Felt like I lost her again. Don't know how many more losses I can handle - R.B.*

During their time in France, the members of the 597th may have been enjoying the pleasures of French hospitality, but not the mission assignment. Even after being banged up by the New Year's attack, the château that had become their home was still a welcomed change from the Nissen huts that served as their quarters back in Abington, but the AR missions were really starting to wear on the men and tensions and tempers were mounting. The Mustangs were proving to be far too vulnerable at low altitude. Two more had been damaged that day and Lt. Pearson had glass fragments in his eye from a shattered canopy. The job was getting done, but at too high a cost of men and materiel. Joe returned from that day's mission tried and sore.

Upon entering his room, he found a letter addressed to him, this time in his mother's handwriting. Removing his boots and jacket, he opened the envelope and sat down in the chair by the fireplace. Inside was a newspaper clipping from the Sharpsburg Gazette. The headline read, "Local girl makes good in a man's world" along with a photo. Joe read on:

"Sharpsburg resident Maureen Fowler, *shown above*, poses with machine shop department head Douglas

## Robert Brun

Hammersmith after presenting Miss Fowler with he Hammersmith Foundry's *Employee of the Year"* Award.

Joe stared at the photograph. Maureen, smiling and looking beautiful was standing next to a tall, strikingly handsome man who had his arm around her waist. Attached to the article with a paperclip was a note from his mother that read,
"Thought you might like to see what Maureen's been up to. We're all so proud of her."
Joe felt his blood begin to boil. So this was *Doug*!
He slumped in his chair. After a few moments, he sat up grabbing fountain pen and paper and exploiting the rage he felt, Joe wrote, the words spilling out effortlessly.

*Dear Maureen,*
*Congratulations on the newspaper article. It sounds like you and* Doug *are having a great time together back there in Atlanta! I wish I could say the same thing is true for me here. The weather is* fantastic! *When it's not snowing, it's raining. Mud is everywhere, and when it's not doing either, I'm out on a mission getting shot at while my girl(?) is back home running around with some Romeo 4-F!*
*So keep up the good work there with "Doug" and the rest of his "Kittens" and maybe you can see clear to fit me into your schedule if I make it home... alive!*

Joe signed the letter *"Captain Joseph Dyer, USAAF"* and as if to add insult to injury, cut his lip while sealing the envelope. Still steamed, he pulled on his jacket and blowing past Rob on the stairs without so much as a word, stormed over to the base office, shoving the letter through the mail slot with more force than necessary and jamming his finger in doing so. Despite the pain, he actually felt better.

The next morning after a restless night, Joe entered the equipment hut and was surprised to see Browning, Taylor and Harrison's replacement, Lt. Mike Wilson gathered over in one corner of the room. Taylor had on some kind of strange looking trousers and Wilson was holding up a matching pair. Walking over, Joe could see laces and a number of tubes and hoses leading out from the suit. Taylor looked very uncomfortable.
"What the hell is that?" Joe asked.
"It's called an *`anti-g-suit.'*" Taylor replied standing up in a rather stiff manner and looking slightly embarrassed. "It's supposed to add 1-g to the number of *gees* a pilot can withstand before blacking out."

## My Shadow

"Look pretty dammed uncomfortable to me." Joe said. "How's it supposed to work?"

Rob spoke up, reading from the Army Instruction Manual that came with the suit.

"Says here:

`...The tubes hook up to the Mustang's vacuum system and the garment is activated by a g-sensing switch that inflates air bladders located in key places throughout the suit that squeeze the blood back up to the head...'

Bad news for you Taylor, your head's already too big as it is." Rob cracked.

"Say what you want Browning, but if it gives me an edge over Jerry pilots, it'll be worth the discomfort. By the way they're three more of these things over in the corner, they've been issued to all the pilots with orders to use `em. Maintenance is making the necessary modifications to the planes as we speak."

"Looks like a dammed corset." Rob groused picking up a set and holding it against himself for size.

The following morning, four rather stiff looking pilots of Red Flight made their way to the flight line and their newly modified planes.

Later that morning on another AR mission cruising at 3,500 feet, Red Flight scanned the area around Marne, France. Since the breakout from Normandy, the allied troops had been pushing the Germans back across France with devastating losses on both sides.

Having their airfields over-run and with the failure of *Bodenplatte* the *Luftwaffe* had withdrawn all of their remaining fighters to Germany. There had been very little activity on the ground or in the air so it came as a surprise when Joe heard Taylor break radio silence and call out four *"bogies"* approaching from five o'clock high.

A quick glance over his shoulder and Joe recognized the four German 109s that now had the drop on Red Flight. Giving the order to break, the Mustangs turned to face the oncoming attackers.

Head to head, the eight planes passed each other without firing then the 109s broke hard left.

"Red flight, they're trying to get behind us. Everyone break left, lets head em off."

Anticipating the German's strategy, the four planes of Red Flight turned hard to meet their opponents and the eight planes forming a circle 3,000 feet above the ground.

Joe, followed by Rob, then Taylor and Wilson made two complete rotations without either group gaining on the other. Tactical options began racing through Joe's mind when he saw the lead 109 break the circle and dive for the deck. He was immediately followed by the three remaining planes all splitting up left, right and up like the petals of a blossoming flower.

"Red three and Four, follow left, Rob, you and I'll take the top, let the other two go, but watch your six-o'clock in case those two return."

Joe pulled back hard on the stick and was surprised by the tight feeling he experienced in his legs. The G-suit had been activated by the force of the sharp climb and was squeezing his lower extremities. The odd sensation was uncomfortable, but this was no time to think about it, there would be time for contemplation later... he hoped.

The 109 continued in a long climbing arc and Joe followed until both planes were pointing almost straight up. Watching the airspeed indicator unwind like a clock running backwards, the Mustang fought gravity, as velocity dropped away. Then the fighter began to shudder, the first sign that he was loosing the precious lift required to keep the plane aloft. A thousand options came and went, none of them any good.

Out in front, the 109 rolled onto its back, pulling hard for the earth three miles below, twisting and turning as it did so. With black smoke emanating from the exhaust ports, the Messerschmitt raced straight down pulling away while *MoJo II* tried to follow. Regaining air speed, control return to the flight surfaces as Joe tried to bring his guns to bear. The Mustang's Merlin was at full emergency power, pulling maximum manifold pressure while in pursuit of the fleeing enemy fighter and closing the distance. Joe had no chance to fire before the Messerschmitt, once again pulled back hard and zoomed into another steepening climb.

"Here we go again." Joe thought preparing to follow the 109 through the same maneuver while racking his brain for any options that would prevent a similar outcome.

Again Joe felt the strange squeezing sensation in his legs indicating the G-suit's operation, but this time there was no question about it, this was painful! Joe could sense the blood draining from his head as the scene around him began loosing color, the g-forces pulling his right hand off the grip and down into his lap along the shaft of the control column. His arms felt like lead and only the Mustangs arm rest and a tight grip kept his left hand on the throttle lever.

Joe was grunting like an animal, fighting his artificial weight and forcing his chin up off his chest when he saw the 109 flatten out before him into level flight.

During his own personal struggle the same forces plaguing him were also affecting the German pilot. Uncomfortable as it was, the G-suit had worked, he had remained conscious; the German pilot had not. With a twinge of guilt, Joe fired a long burst and watched the 109 turn over and spin into the ground.

# My Shadow

Readjusting his oxygen mask, Joe wiped the sweat away from his eyes feeling the bladders in the G-suit relax allowing the blood to flow back into his legs. Joe smiled, looking down at the strange garment with a new appreciation.

"Mad-dog Red two to Red leader, come in. Over." Rob's voice broke the moment. "You okay Georgia?"

Out of breath, Joe replied. "Red leader to... Red Two. A little shaken, but okay."

"That was some roller coaster ride you took Georgia. I could hardly keep up, and what was all the noise about?" Rob asked once he'd pulled up alongside his flight leader.

"What do you mean?" Joe asked.

"I mean all that grunting and groaning. Sounded like a bad case of constipation. You need Doc Lewis to prescribe a strong laxative or something?"

Confused, then embarrassed, it occurred to him that all during the excitement of the battle, he'd had his thumb pressed firmly down on the radio transmit button atop the throttle control lever. He had broadcast the entire engagement over the radio.

"Six and a half Gees." was all Joe said.

Back at the base, Joe taxied *MoJo II* over to the hardstand and shutdown the engine. Exhausted, he slid back the canopy and looked over to see Taylor and Wilson's standing nose-to-nose and yelling about something. Joe climbed out onto the wing, jumped down and trotted over to where the two men stood arguing.

"Well from now on just stay out of my way!" Taylor shouted red faced with anger.

"Stay out of YOUR way?" Wilson screamed back. "You ought to consider yourself dammed lucky I was *IN* your way or that *Kraut* pilot would have had your sorry ass splattered all over the Marne."

"Are you kidding? Two more turns and I'd have had the Hun bastard cold!"

## Robert Brun

"In a pigs eye!" Wilson replied.

Joe had seen the men wound up after a mission before, but this appeared like the two men were going to come to blows. He was about to intervene when Lt. Anderson, the bases Adjutant, rushed out onto the field.

"Guys... Guys!!!" Anderson called, out of breath. "I just got word. We've been relieved, we're pulling out. We're going back to Abington!"

With this news, a silence fell over the group and both men unclenched their fists. Taylor slumped down onto the bumper of the jeep while Rob, who'd been silent throughout the altercation, took a deep breath and sighed.

"I'm sure gonna miss this place." Rob said then turned and wandered off toward his plane.

The Armed Reconnaissance flights that he and the men had been flying had never sat well with anyone. Naturally nerves were on edge, but he'd never seen Taylor and Wilson behave like this before. He'd be glad to get back to Abington. The men's stress levels, after so many months of close ground support, had finally reached the breaking point. It was uncomfortable being this close to the action or the ground. It would be a relief to escort the Big Friends once again.

Joe found Rob standing by his plane and called over to him.

"So we're heading back to England. I guess with Command ramping up the bombing campaign, our Big Friends will need all the help they can get. Leave the *Jugs* to clean up the rest of things here."

"Yeah, clean up." Rob relied not looking over. "Look Georgia, I got a few things to take care of before we leave. I'll catch up with you at chow."

Later that afternoon while in his room, Joe was packing the rest of his gear when Rob came in looking confused and frustrated.

"What's up my friend?"

Without saying a word, Rob slumped down in an armchair in the corner staring out at nothing in particular.

"Joe... I can't go with you... I mean, not yet!"

"What? What do you mean you can't go? We've got orders to return to Abington and that means everyone, including you."

"I know, but there's some unfinished business I have to take care of before I leave and I don't think I can do it before tomorrow."

"What are you talking about? You haven't gone and gotten some *Mademoiselle* in trouble, have you?"

"No, nothing like that! It's just that... I can't really say. It's, well... kind of a secret and all."

"Rob, just what the hell are you talking about?"

## My Shadow

After a moment, Rob spoke up. "Oh al-right, I'll show you. Come on with me."

Rob led Joe over to motor pool and climbed into one of the jeeps parked there.

"Get in," Rob instructed.

The two men rode in silence out the dirt road that exited the vineyard and ten minutes later entered the small village just outside of Sarrebourg. Everyone there seemed to recognize Rob as he maneuvered the jeep around the cratered main road through the village. Down toward the far side of the village stood a small stone church with one section of the roof caved in and another under repair.

Rob stopped the jeep and without a word showed Joe into the church where they met by an elderly priest who spoke very little English. Rob greeted the holy man with a kiss on both cheeks and proceeded to lead the way into the sacristy where Joe's eyes beheld a beautiful sight. There, on the wall behind the altar was a magnificent mural depicting the ascension of Christ. Joe let out a loud whistle then remembering where he was, caught himself with a cough. The elderly priest began to chuckle.

"You did this?" Joe asked looking at Rob

"Well, not exactly. The 17th century fresco was badly damaged during the liberation and I'm helping put it back together. It was our artillery that put it in this state. I just figured it was the least I could do. Now you see why I can't leave. Not yet." Rob said.

"I've been working on it off and on since our arrival and it's almost done, but I'll need about another day and a half to finish up. You guys can fly on ahead, I'll keep *Shadow* here for a day or two with...`engine trouble'* and be back in Abington before the end of the following day. What-da-you-say?"

"Tomlinson's not going to like this..." Joe said then his eyes met those of the priest. "...But I'll see what I can do."

Before Joe could say another word, the priest had him by the hand and was shaking it gratefully, kissing Joe in gratitude and jabbering rapid fire in French.

"Yeah, yeah, yeah," Joe said wiping his face with his handkerchief. "You can thank us both in the stockade."

The next morning, Red Flight minus one Mustang, took off for the flight back to Abington. *Shadow II* had developed `ignition trouble' and was to follow a day later.

## My Shadow

# Part IV
### Chapter 28
### Trains

*Tuesday, February 6, 1945: Something's definitely up with Georgia. Practically ran me over the other day in a real hurry. Better keep an eye on him for a while. We're back in England though and back upstairs. Not a minute too soon. That ground war really stinks - R.B.*

The flight back to Abington was uneventful and the homecoming was less than enthusiastic. In the six months since they had been stationed in France, the 597th's remaining fighters had flown over two hundred escort sorties with the loss of fifteen pilots. Many of the men Joe had served with before the transfer to France were now gone, their replacements all strangers to him.

It was with an odd sense of *deja vu* that Joe reentered the Nissen hut and tossed his pack down on the cot he'd vacated six months earlier. There was an eerie silence in the building. Cushman was dead, Adams was dead, Davis captured, Harrison dead and others that had come and gone since he was away that he'd never even met. Joe was experiencing the same sense of isolation and disorientation he'd felt his first day at Basic Training, but without a trace of the enthusiasm, just the loneliness.

Joe didn't have much time to dwell on this feeling though because two days later he was back in the air leading the squadron and catching up to the bombers again. It was good to be back at altitude, although a great deal had changed.

From where he flew, Joe saw the reflective glint of a multitude of unpainted B-17. Like his Mustang, the newer B-17s no longer wore the familiar olive drab paint and the sunlight shown off their mirror-like surfaces. There also appeared to be a complete absence of enemy fighters.

Joe scanned the horizon for telltale traces of bogies, but saw nothing. Nothing, that is, but the hundreds of escort fighters zigzagging their way along the contrails. He hadn't seen so many planes in the sky at one time since D-day.

Still, flak over the target, where no amount of fighter cover could make a difference, was heavy and Fortresses continued to be hit during the bomb run.

Shortly after turning for home, the *Luftwaffe* once again appeared, but these enemy fighters made only a half-hearted effort to break up the flight. Low on fuel, ammunition or morale, the Mustangs easily ran them off.

Nearing the French border, and with no further sign of enemy fighters, it was time for a little free-lance work. Although officially back from A.R. in France, it was now the standing order that once relieved by the replacement escort, all fighters were expected to seek out ground targets over the Continent.

"Anyone with left over ammunition is now free to engage targets of opportunity," came the general call over the radio.

"Thanks for the help, Little Friends." came a reply from the bomber's squadron leader. The Thunderbolts of the 358th having taken over escort duty while Joe and the Mustangs of Red Flight peeled off from the formation and headed down to the deck for a look around.

"Mad-dog Leader to Red Flight. Wilson you and Taylor, head east and check out sector 213. Red Two and I will head north."

"Roger Mad-dog Leader," came the reply, and two of the Mustangs broke right while Joe and Rob formed up heading north towards the village of Rouen.

Circling in a descending spiral, they leveled off at 1,500 feet and began scanning the horizon. The Germans were still trying to bring down troops and equipment from the north to supplement their forces that were taking a real shellacking in the south as the allies continued their advance inland toward the German border.

"Why did they continue to fight?" Joe wondered.

Turning his head to the left, Joe spotted a plume of black smoke rising up from the tree line about five miles away.

"Looks like *Fritz* is on the move again." Joe said keying his mic."

"I see it." Replied Rob. "Let's go get some Chattanooga choo-choo." and both planes simultaneously broke right.

Moments later, approaching the rails from 11 o'clock at 200 yards, Joe opened fire, raking the locomotive with a spray of 50 caliber bullets. Banking off to the left, clouds of steam and smoke erupted from the engine's ruptured boiler while the train slowed to a halt.

Whirling around for another pass, the two Mustangs regrouped in a string formation, one after the other, this time lining up with the tracks and strafing the train from engine to caboose. As they passed over the

## My Shadow

center of the train, the top of one of the boxcars open and two quad 20mm anti-aircraft guns opened fire on them.

"No fair shooting back." Rob said sarcastically into his radio, tracer rounds flying passed his canopy.

"Better break off " Joe suggested. "Getting a bit hot around here, besides, I'm out of ammo."

"I got just enough for one more pass." Rob announced. "You start back, I'll catch up."

"Roger, but don't dally none, those quads mean business."

Joe climbed his plane up to 2,000 feet and looked back while his wingman circled around for one final run. Rob began his pass and Joe's concern grew when the twin quads open fire at the oncoming plane. Although out of ammunition, Joe was unable to simply sit back and watch, so dropping his starboard wing, he pressed the Mustang over into a screaming dive, the sound of his Merlin reaching a high-pitched crescendo.

It had the desired effect. The two anti-aircraft crews scrambled to swing their guns around believing they were about to be attacked from both sides. During the panic that ensued, several of the German soldiers dove from the platform of the now stationary train and ran for cover in the nearby trees.

Joe pulled back on his stick, leveling off just above the treetops and turned to see his wingman, machine guns blazing, pull hard and pass a few feet from his left wingtip as the adjoining boxcar exploded, the train rearing up like a gigantic inchworm.

Climbing into a tight spiral, Rob leveled off, inverted, then rolled his plane back over while Joe executed an Immelmann and pulled up alongside. Below were the remnants of the smoldering boxcar, clouds of steam rising from what had been the train's boiler.

"Wow! Georgia, that was sure a close one. Next time I'll take your advice. Thanks for the covering fire! Over."

"Anytime, buddy, but let's keep that kind of maneuvering limited from now on, huh? I think I lost a few feathers on that last pass. Over."

"Oh, you never let me have any fun. Out" Rob whined, the two Mustangs joining up and headed back toward the Channel.

The countryside below was bucolic and Joe was actually able to relax a bit. Over the next half hour, he enjoyed a pleasant, low-level flight without the threat of enemy fighters or flak. It was just like it had been earlier. Feeling as though he'd done his part for the day, Joe eased back into his seat.

Fifteen minutes later with the French coastline in sight, Joe spotted something streaming from his wingman's plane, a faint mist of liquid trailing off from the cooling-air exit flap below the fuselage.

## Robert Brun

"Hey, Rob," Joe called, "I think you've got a coolant leak. Hold her steady... I'll take a look."

"Roger, Cap."

Joe dropped back and below Rob's plane for a better vantage point. There was no doubt about it he could see a thin line of fluid streaming out from the radiator vent. It wasn't much, but it was there.

"Rob." Joe called back. Looks like you took a hit in the radiator. How's your engine temperature? Over."

After a moment came Rob's reply.

"Temperature reads 20 above normal and climbing slow, but steady, nothing too serious yet. Over."

"Good, but you're loosing coolant. You must have taken a hit on that last pass. Keep an eye on your temperature gauge and keep me posted. I'll radio ahead on the emergency channel for clearance. See if we can make it to the RAF base at Dover."

For the next forty-five minutes, Joe kept an eye on the coolant leak from his wingman's plane. Every ten minutes, Rob would call over and report the reading on the temperature gauge as it continued to climb. How long did they have left? Minutes? Seconds? The cold, dark grey waters of the English Channel looked far from inviting. Joe could imagine the pistons of Rob's engine grinding into the cylinder walls of the overheating Merlin.

After what seemed an eternity, the Dover coastline finally appeared and Joe called the base on the emergency frequency once again. Seconds later, they spotted the airfield and received clearance to land.

"Okay, Rob," Joe called. "You're cleared. Take her on in."

"Rog...!" and no sooner had Rob lowered the gear than the Merlin's propeller spun to a stop.

"Well." Rob called back. "See you on the ground. Out."

Joe pulled his plane out of the way and positioned himself above and behind his wingman. He watched while Rob dropped the flaps in preparation for a "dead-stick" landing.

It was a strange sight to see the Mustang in the air, flying along beside him without a spinning propeller. Checking his air speed, Joe was dangerously close to a stall as the two planes mushed along. It was no longer possible to remain along side, so advancing the throttle Joe broke formation with a salute to Rob.

"You'll have to take in from here." Joe said

"Beggin the Captain's pardon," Rob called back, "but I'm kinda busy right at the moment."

While pulling ahead, Rob's plane sideslip into position on the glide path, flare out at the very last instant just inches above the field, and

## My Shadow

drop onto the ground with a small bouncing roll before coming to a complete stop.

Even before Rob was able to get the canopy open, four RAF ground crewmen hopped out of an approaching truck and were already pushing the plane off the runway. Joe positioned his plane in preparation for landing when Rob called on the radio imitating a heavy British accent.

"*Beg Pardon Gov-nah, but they tell me they av a flight of `Lancs' inbound that are in desperate need of this field and it seems you're in the way. Would you be ever so kind as to give them a bit more room? Thanks awf-ly. Over.*"

Looking behind, Joe saw about a dozen RAF Lancaster bombers, many showing signs of battle-damage, queuing up in preparation for landing.

Without bothering to reply, he broke off his approach, raised the wheels and flaps and banked sharply off to the right climbing out of the way of the descending bombers.

"Listen, Georgia, there's nothing more you can do here. Why don't you head on back to base and fill them in on my where abouts. I'll catch up with you as soon as I have some *tea and crumpets with these Brits*. Old Tomlinson must be worried sick about you by now. Over."

"Roger. Both of us, I'd guess." Joe replied. "Just make it a short one. Don't forget, we're scheduled to fly again tomorrow and you know I don't go anywhere without my *Shadow*. Mad-dog out." Joe rocked his wings and headed off back toward Abington.

A half hour later, Joe felt the reassuring rumble of his wheels touch down back at the 597th's field. Sliding back the canopy he taxied the Mustang over to the hardstand where Zeke flagged him in and gave the signal to shut down the engine. The first words out of his mouth were inquiring about Rob.

"Coolant leak." Joe shouted in reply clearing his ears and adjusting to the silence. Had to put down at the RAF base at Dover, but he should be back by tomorrow, next day the latest."

## Robert Brun

"I'll inform the Colonel." Zeke remarked hopping up on the wing and unfastening the harness. "Any trophies today?"

"One in the air, but I also got a locomotive on the way back."

"Good, that ought to mess up that line for a few days at least. I'll see to it *MoJo* gets credit for both."

"Thanks, Zeke." and Joe jumped down from the wing heading back toward the hut. "Keep me posted on Rob, will ya?"

After chow that evening at the Officer's Club, Joe entered, surprised to see Rob sitting and chatting with Lt. Taylor. Approaching the table, Browning and Taylor looked up at Joe.

"Rob?" Joe inquired taking a seat at the table. "How on Earth did you get back here? I wasn't expecting to see you until at least tomorrow."

"He landed about half an hour ago." Taylor announced. "Flew up from Dover."

"Flew? I thought your engine was shot! I saw it seize up with my own eyes."

"I shut her down." Rob said somewhat sheepishly.

"YOU WHAT?" Joe said in almost a shout.

"I shut her down. The engine temperature kept climbing and I knew that if I let it go much longer, the whole thing would seize, so once I got the base in sight with the wheels locked, I shut the engine down. Do you have any idea what one of those Merlins cost the taxpayers?

"It turned out that all I had was a small break in a secondary coolant hose - one of those quads must have nicked the line. So happens the chief mechanic at the base used to work on Mustangs when the Brits first got the `A' models and spotted the trouble right off. He patched the hose with a tin can, piece of hose and some clamps, topped off the coolant and by the time I'd had a *cuppa* and a *bickie*, I was on my way. *Bloody nice chaps that lot, aey.*"

Joe just stood there blinking in silence.

# My Shadow

## Chapter 29
## Saint Valentine's Day Massacre

*Tuesday, February 13, 1945: It's good to be back upstairs again, but I'm worried about Georgia. He's turning into a real "Mother Hen" with the flight. Always seems on edge. The guy worries too much - R.B.*

Just after midnight, on the morning of Wednesday, February 14, 1945, Joe found himself lying in bed, staring at the ceiling and wondering if he even had a *Valentine*. He sure didn't expect to be getting any cards in the mail, not after what he'd done.

After the letter he'd sent, he'd be surprised if Mo would ever speak to him again much less have any feelings toward him, other than anger. How could he have been so stupid? There was no other word for it. He'd been mad and he had been jealous of *Doug,* but now, in retrospect, he didn't even know what, if anything, there was to be upset about. So what if he'd taken her dancing, what was she suppose to do while he was away, sit quietly staring out of her bedroom window pining for her *hero's* return? He should be proud that she had been made 'Employee of the Year,' everyone else back home was. Joe felt childish.

"No news is good news, I guess." he told himself, but that sure wasn't the way it felt. It still aggravated him not to know what feelings Mo might have for *Doug* almost half way around the world while he continued to put his life on the line day after day.

Sure, he hadn't heard from Mo as often as he'd liked, but what she had written, when Joe took the time to really stop and think about it rationally, it was no '*Dear John'* letter or a casual friendship. Maybe she really did love him as much as he did her and that's what made it so hard to

## Robert Brun

understand. Deep down inside he simply couldn't figure out why Mo would love him, after all, who was he?

Oh sure, he was a pilot, a *fighter* pilot, a Squadron Leader and an *Ace*. He was one of those suave dashing flyers you see in the movies and a Captain in the USAAF, fighting the Nazi menace to make the world safe for democracy, but what had he really accomplished, and what did any of it mean in the long run?

It didn't take a military genius to know that it was only a matter of time before the war would be over. The Italians had already surrendered and the allies were pushing hard from the west while the Russians had crossed the Oder River. Why the Nazis continued to fight was a mystery to Joe, but they did and therefore, so did he. Everyone knew the end was near and perhaps that's what really worried Joe the most. He'd be 22 in another two months and all he'd ever done was milk cows, fly planes and...
... And <u>kill</u> people!

This realization surprised him. He tried to push it out of his mind, but it wouldn't leave. He was a killer and had to admit it. A hired gunslinger, just like in the dime store novels he'd read as a kid, but this wasn't man-to-man like the *Mexican Standoffs* where one gunfighter faced another at high noon. In aerial combat you snuck up on the guy from behind, and shot him in the back when he wasn't looking, hardly honorable. Even so, this was the way it was, and this too was only temporary. What was he going to do afterward, assuming he'd make it? Joe hadn't a clue and this worried him the most. Doug was handsome, a Foreman at a foundry, he had a trade, a career, he could provide. Hell, even Mo had more work experience than he did. Joe was what, a fighter jockey? That an ten cents got you a cup of coffee.

"Rise and shine, gentlemen." The Corporal's shrill voice rang out rousting the men in the other sections of the hut. Joe sat up in his bunk and kicked Taylor's cot with his foot jarring the man awake.

"Saddle up, Danny boy... it's time we ride."

Taylor looked puzzled by the cowboy reference, but grumbled just the same. Lt. Taylor was not a morning person.

As Squadron Commander, Joe had already seen the mission roster and knew he would be flying that day though as yet, he didn't know where the mission would take him. It really didn't matter, another day, another German city. Joe couldn't have been more wrong.

The city of Dresden lay in the eastern corner of Germany, eighty-five miles south of Berlin and close to the Czech border. As the Capital of Saxony, the city had been a cultural center for well over a hundred years. In the past year it had also become a major transportation hub and virtually all German troops and supplies going to and from the Eastern Front had to

## My Shadow

go through Dresden. It had also become filled with refugees fleeing the advancing Red Army.

This German city had been bombed several times before, but so far, only as a secondary target. Despite containing several military related industries including a small oil refinery, it had not shared the same strategic significance that cities like Hamburg or Leipzig had up to this point. Now, with the Russian army closing in just eighty miles to the east, Stalin was screaming about Dresden and threatened to withdraw his troops if something wasn't done. As a result these marshaling yards had become the primary target this day.

At that morning's briefing, Joe learned that the RAF had already hit Dresden twice the night before and the 8th was to be the third wave over the target around noon.

"Three raids in one day?" Joe puzzled; it seemed excessive.

Escorting the B-17 bombers to the Czech border would make for a very long day, so Joe decided to lay off the coffee this time.

The 316 Fortresses stretched out far ahead of the fighters while Joe and Rob followed the rest of the escorts half way across the European continent. For the nine months since the Normandy invasion, the resistance of the *Wehrmacht* had continued to weaken while Generals Patton and Marshall pressed hard from the west and General Zukov, leading the Red Army, pushed in from the east. Aside from the Ardenne offensive, which was now being referred to as the "Battle of the Bulge," and the *Luftwaffe* attacks on New Year's Day, German resistance had all but collapsed.

Joe had been flying for over four hours and had yet to see an enemy plane. He flexed his fingers while switching hands on the control stick. Damn it was cold!

Constantly scanning the sky despite the complete lack of opposition, Joe could see smoke way off in the distance, lots of it. A huge column of thick, black smoke rose like a fountain along the horizon. Joe checked his map confused, and then checked it again. According to his information, the IP was still over one hundred miles away and yet, there the smoke was.

For the next thirty minutes, the smoke column grew in size until the bombers finally reached the outskirts of the city. Below, there was nothing to be seen, but smoke and flames. It appeared as though the entire city of Dresden was burning.

With nothing to do, but stay out of the way, Joe watched the bombers released their loads into the conflagration. It was like throwing a match into a forest fire; adding insult to injury. Had things really gotten to this point?

## Robert Brun

Regrouping with the bombers after their turn for home, nothing below could be seen that indicated they had been over a city. Even the usual flak was weak and ineffective, most of the cities anti-aircraft defenses having been moved east as anti-tank guns against the advancing Russian Army. The radio too was unusually quiet for the return trip, devoid of the usual chatter between planes.

Hours later, after landing back at Abington, Joe made it through the debriefing with a strange sensation of disapproval, like he'd just done something wrong. Major Nealsen seemed to sense it too, for his questions were less probing than usual.

In the equipment hut, Joe sank onto the bench with the full weight of his body and put his head in his hands. He was exhausted. The rest of the pilots began to enter, but conversation was still at a minimum and what there was of it dealt with plans for later that evening. Lastly Rob entered and storing his gear, glanced over at where Joe was sitting. Joe looked up and met Rob's gaze.

"Damn!" Rob said, turned and walked out of the hut.

"You said it, buddy." Joe said to no one in return.

# My Shadow

## Chapter 30
## A New Threat

*Thursday, April 5, 1945: Dresden was really bad! Never seen anything like it before. I sure hope this is over soon. I'd really like to go home - R.B.*

Joe had heard rumors of a new type of German fighter, but hadn't given it much thought. He already had his hands full with the Messerschmitts and Focke-Wulfs that he had been dealing with up till now.

"*Jets,*" they were being called. Planes that flew without propellers and were supposed to be fast, was about all he knew of them. So it didn't come as a total surprise when, two swept-winged aircraft approached from above and behind at extreme speed. The contrails confirmed they were very fast and approaching from high above the bombers.

"Mad-dog leader to squadron. Gentleman, we've got jets." was all the warning Joe gave to the rest of the squadron watching the strange aircraft slashing their way through the Allied fighter cover and into the bomber formation.

"Red two, this is Mad-dog leader, Rob, follow me. Over."

"Roger"

The two planes took off in pursuit, trying hard to keep an eye on the speeding aircraft. Joe firewalled the throttle pulling 60 inches of manifold pressure and split-Sed, putting the Mustang in a steep dive in pursuit of these strange, oddly shaped aircraft. None of the rumors he'd heard, however, had prepared him for the event unfolding before him.

The two dark silhouettes looked more like sharks than any aircraft Joe had seen before. The wings on these planes were swept back, giving them a distinct inverted V-shaped appearance. Below each wing a tubular pod emitted a faint stream of smoke that Joe correctly assumed to be propelling the aircraft.

The jets leveled off behind the bombers, pulling up from below to bleed off the tremendous speed they had developed in their powered dive. Now appearing to be almost stationery, they floated there just beyond the range of the bomber's guns.

Joe closed in on the aircraft, but before he was within range, multiple rows of smoke trails shot out from beneath the wings. Seconds later, two of the fortresses exploded, falling out of the formation. Realizing they'd been seen, the jets then dove, accelerating rapidly and pulling away too fast for the Mustangs to pursue.

"Let em go, Red two." Joe called over to Rob retarding the throttle and watching the four smoke trails disappear into a nearby cloudbank. "We'll never catch 'em at that speed. Head back to the bombers." The two fighters rejoined the bomber group.

The rest of the mission proceeded without further fighter encounter, instead, Joe again sat by and watch the deadly accurate German flak blackened the sky over the target. In his headphones, he could hear the constant drone of the German tracking radar and witnessed two more fortresses hit and fall from formation. He no longer looked for chutes

Regrouping with the remaining bombers at the rendezvous point, the trip back to the Dutch coast was quiet. Switching over to C-channel, the bomber's frequency, Joe radioed the group leader.

"Big Friends, this is Mad-dog Leader. Looks like you can make it from here. I think we'll linger for a bit of sight seeing. Over."

"Thanks, Little Friends. Give em a few rounds for me, and the rest of the boys. Big Friends out."

Joe rocked his wings in salute, he and his wingman then pulling up and away from the bomber stream.

Switching back to the fighter frequency, Joe called over to Rob.
"Mad-dog leader to Red two, how'r you set on fuel? Over."

## My Shadow

"Got enough for a look around, Georgia." came Rob's reply.

"Good. Let's circle back, I want to check out something I saw a ways back. Over."

"Got you covered, Mad-dog leader."

The two planes banked around, Joe scanning the horizon for the light patch of ground he'd seen on their way to the coast. It was just a hunch, he knew, but he had enough ammunition and fuel left for a quick look around.

Not five minutes passed before Joe caught sight of what he'd been looking for. Down below him blending with the trees and heading in the opposite direction was the same type of swept-winged jet aircraft he'd seen earlier that day.

"Can you see that?" Joe radioed to Rob.

"Yeah, I see it, and his wingman too."

"Follow my lead."

"Don't I always?" Rob called back with a faint trace of sarcasm.

Banking, the two Mustangs circled the unsuspecting jets by pulling around and placing the sun behind them. Below, the jets flew on preparing to land, their flaps extended and landing gear down. Out ahead, a narrow break in the trees showed a long stretch of roadway beside a small, hidden airfield.

Throttling back to avoid overrunning the landing jets, he lined up his sight and opened fire at 300 yards. Immediately, the left engine of the rear jet caught fire and the jet exploded in a red-orange ball of flame.

Noticing his comrade's predicament, the remaining jet responded with evasive action. While the Mustang pilots watched, two long tongues of flame shot out of the engine exhausts and just as quickly disappeared. An instant later, the jet pulled up into a steep stall, rolled over onto its back and crashed into the ground.

"What happened?" Joe said keying his mic.

"Like shooting rats in a barn. You must have scared that second one to death, Georgia." Rob called back seeing what had just happened. "I don't know if that one counts? Over.

"What do you say we get out of here?" Rob asked.

## Robert Brun

"Not with all this leftover ammo," came Joe's reply

"Roger" and the two pilots leveled off, lining up with the runway and strafed the length of the field, shooting up planes, buildings and vehicles along the way.

Once at the far end of the field, the two planes split up and climbed for altitude, escaping the burning planes and vehicles below before any return anti-aircraft fire could get their altitude and range. Going back for a second pass, although tempting, was always dangerous.

Climbing to 8,000 feet, Rob pulled up alongside Joe.

"That'll make for a pretty drafty barracks tonight. Maybe it'll give 'em all a case of pneumonia?"

"We all do our part."

After returning to the base during the debriefing, Joe filled Major Nealsen in on the details of what he and Rob had seen. The major was very interested in their observations of these jet aircraft.

"So the second plane spouted a jet of flame and then crashed?" The major said looking up from his clipboard.

"That's about it." Joe answered. "He must have applied power once he saw we were behind him and it was as though his engines just... ...blew out!"

"Hmmm..." was all Nealsen said while scribbling away frantically on his notepad. There was a long pause and then he spoke up again.

"Was there anything else, Captain... Lieutenant?" he asked looking up at the two men. This time it was Rob's turn to speak.

"Well, I did notice one thing. Despite their speed, they didn't appear to be all that maneuverable. I mean, their con-trails all seemed pretty straight to me. I'd bet I could turn inside any one of 'em. I also noticed that they seemed to attack the bombers from below and behind - not the frontal attacks we're used to."

"Most likely because of too great a closing speed head to head." Major Nealsen replied.

"They seem to dive down from above the fighter cover too fast for anyone to catch and then let the speed bleed off, slowing down quite a bit after they dipped down below the stream. Now my guess would be, if we were to position a few fighters ready to pounce just behind the bombers, those jets would be sitting ducks. There was another pause while the major continued to take notes, then Nealsen spoke again.

"Thank you gentlemen, you've been most helpful." The major said collecting his things and starting to leave.

"One more thing, gentlemen… keep this to yourselves." and then more sternly, "That's an order!"

Joe and Rob stood there silent.

## My Shadow

"Keep it to ourselves!?" Joe blurted out while stowing his equipment. "Like hell I will! Information like that would keep our guys alive. They have a right to know."

"You're right Joe," Rob said trying to calm his Flight Leader down. " But it's not our job to figure all this out, our job is just to shoot down the damn enemy planes."

"I know, but this is different... Isn't it?"

"Oh, I don't know?" Rob said exasperated, "but you're not going to do anybody any good in the stockade, especially me."

Joe threw his flight helmet against the locker and mumbled something unintelligible under his breath.

"Look, it's been a long day, what do you say we head on into town for a change of scenery and relax? I'll buy."

None of the motor pool bikes were available that evening so Joe and Rob walked to the village. The three miles gave Joe a chance to calm down and reflect on the past few days. He hated to admit it, but he had been on edge since sending his last letter to Mo. He was also steamed at Major Nealson's order to keep the jet information quiet. It made no sense to him. Less and less did.

Chatting with Rob while walking into town, it occurred to Joe that, with all the activity, he had completely forgotten, today, April 5th was his birthday. 22 years old and his second birthday since leaving home, Joe wondered how many more would be spent overseas? Not feeling much like celebrating, he decided to keep the information to himself.

Entering the pub, the two airmen gave a series of perfunctory greetings to the people there and made their way over to the bar. Unlike other evenings, the pub seemed more solemn tonight and Joe noticed how many new faces there were. Had that many airmen been lost since his arrival at Abington? Still upset, he ordered himself a whiskey.

Looking at the veteran pilots there in the pub, Joe could see the same signs of fatigue on their faces that he too was feeling, too much stress, and too many missions. The pilots had taken on a certain *look* after a few dozen missions. Pale coloring, sunken and mildly crazed eyes that never seemed to settle on one object for more than a second, all were symptoms of the job they performed day after day.

Joe drained his glass and ordered another drink then looked around the pub at the men there. It was easy to distinguish those who had been here awhile from the replacements. There were not many of the original pilots left. Some were dead others were guests of the Germans and still others had just disappeared, but all were gone.

## Robert Brun

Noticing a particularly drained looking face, Joe was surprised to see his own reflection in the bar mirror, staring back at him and looking much older than his 22 years.

"You okay?" came a voice at his shoulder and Joe turned to see Rob looking concerned.

"Yeah... I guess. Just getting a bit stuffy here at the bar." and ordering a third drink, Joe got up and moved to one of the booths over by the door.

A bomber crew from Kimbolton began to sing, loud and in an off-key way that seemed to compliment the rude topic of the particular melody they had chosen. Some things never changed.

Joe extracted the last letter he'd received from Mo and reread it for the fourteenth time still trying to interpret what he hoped was and wasn't there. He hadn't heard back from her since he'd sent that last letter, but then he hadn't written again either, mostly out of embarrassment... and fear. He'd really flown off the handle at her and for what, a stupid newspaper article? Besides, what would he say...? ...Maybe, "*I'm sorry*" for starters? Angry as he had been, he still missed her. All this uncertainty was infuriating.

Reaching into the pocket of his tunic, Joe removed the tattered photo of Mo and sat staring, the image blurring before him.

"Some birthday." Joe thought finishing his third drink.

The room seemed to darken and, looking up from his thoughts, Joe saw the silhouette of his wingman standing next to the table holding two pints of stout.

"Mind if I join you, Captain?" Rob said as he set the two pints of stout on the table. "Here's that drink I promised, but it looks like you've got a head start." noting the empty glass.

"Suit yourself, but I'm not feeling like very good company tonight." Joe said returning the photo to his tunic pocket.

"Girl trouble?" Rob inquired, sliding one of the pints across the table to Joe.

"Yes, and no." Joe said taking the offered glass of dark liquid. "I mean, no *Dear John* if that's what you mean, not yet at least, but yes, because I'm over here, she's back home with some 4-F and I'm not even sure what I'm doing anymore."

"Whad-d ya mean?" Rob said, feeling out his friend, and taking the seat across from him. "We're here making the world safe for dee-MOCK-cra-see." trying, unsuccessfully, to lighten the mood.

"Yeah, yeah. I know all about that!" Joe said waving his hand. "And I should be proud and honored to be participating in stamping out Hitler and the rest of his Nazi thugs who seem, for reasons I can't seem to

## My Shadow

understand, want to take over the world. And maybe that's just it, I can't figure out why! Why on earth would *anyone* want to take over the world?" Rob took a pull on his pint and let out a long sigh.

"That, my friend, is a question best left to Generals and Politicians. I don't know about you, but you could easily put everything I know about world politics into a thimble and still have enough room to float the battleship Arizona, but that's not our job! We're here to do one thing and that is to poke holes in as many of Göring's planes as they throw at us, so those bomber boys, standing over at the piano currently massacring *"White Cliffs of Dover"* can fly up Adolf's shorts, drop their bombs where they'll do the most good, bringing this war to an end so we can all go home, make babies and get on with our lives!

"Look, I know you miss your girl. I can see it on your face every time you pull out that photo or even look at your plane, but just be thankful you got someone home waiting for you, praying for you and looking forward to your return."

Joe wondered if Rob was right. Was Mo really waiting for him and simply killing time with *Doug* or was there more to it? And what right did he have to even expect her to wait for him knowing that any day he could be dead or worse? He didn't have an answer.

"But what if I get killed? What'll happen to her then?"

"Probably run off with some jerk."

Joe shot Rob a nasty look and he realized this was not the proper time for levity.

"Whoa... now hold on there, big fella." Rob said leaning forward across the table and lowering his voice.

"You've got this thing all wrong. Now, I don't know this MauREEN, except for what you've told me, but based on that, she's no shrinking violet. From the sound of things, she's a pistol, a 45 caliber I'd say, and regardless of what happens to you she'll be fine."

"What do you mean by that!" Joe said narrowing his eyes and staring across at Rob. He could feel the alcohol beginning to cloud his head as the image of the newspaper photo of Mo and Doug flashed through his mind.

"Hang on, hang on, save that for the *Brats.* What I'm trying to say is, Joe, you got a hell of a woman there. Strong, loving, caring, kind hearted, hard working, loyal and, for reasons I'll never understand," Rob added sarcastically, "head over heels for you. Someone like her needs someone like you in return. Not to protect her and take care of her. Nah, this is 1945, man. She's back home building the machinery that keeps us going here. Yeah, maybe she hasn't got some *Brat* fighter jockey trying to ventilate her flight suit, but she's just as much a part of this war as you and me, and I'll bet she understands exactly why you're here and the risks we're all taking."

Rob sat back and took another pull on his drink. Starting again, this time sounding more serious than concerned.

"If you start thinking about playing it safe, about keeping yourself alive because you're afraid your girl back home isn't gonna be able to handle it, you're gonna get both of us killed and that's definitely not part of my game plan. Besides, looking after your backside... that's *my* job."

Joe stared at his glass thinking about everything Rob had just said. He knew Rob was right and that's what bothered him the most, because he didn't want to admit it. He was the Squadron Commander, a double Ace Mustang pilot with the United States 8th Air Force. He was a leader of a squadron, he didn't need to be getting advice from his wingman and Joe resented the intrusion, but still he felt lost, he couldn't even manage his own affairs. Some leader.

Joe finished his pint and thought back over the last year, all he'd been through, his successes and failures. His head was starting to spin. It was all getting to be too much and he was tired. Tired of the fighting, tired of leading, tired of seeing his friends disappear, tired of risking other men's lives and worrying who would be next to die. He just wanted it to end, be over, to go home, but to what? Was there anything left there? He'd been away so long, maybe too long.

"Just think about it will ya?" Rob got up and started to leave, then stopped. "By the way, I've been meaning to tell you, your letter came back, `insufficient postage.' Thought you might like it back. Happy Birthday Georgia." and with a wink, he tossed Joe's letter to Mo onto the table.

Joe sat there staring at the envelope suddenly unable to speak. Was it possible? The letter, the source of all his self-torment for the last month was now lying there before him. It hadn't been sent. Mo hadn't received it or read it, but how? Rob again!

Numbed by the liquor, Joe's head began to swim. This was too much to take in at once and as if a dam had burst, months of pent-up anxiety and frustration suddenly spilled forth all at once overwhelming reason. Joe looked up at Rob who stood there with a self-satisfied smirk on his face and something snapped.

"You ga-dam meddling self-righteous BASTARD!" Joe said raising his voice to a shout. Rob stared and the pub fell silent.

"The almighty *SHADOW*!" Joe mocked. "Always following and prying his way into where he doesn't belong. What makes you think you have the right to interfere in my business, my personal LIFE!?"

Rob was stunned; he'd never seen Joe act like this before.

## My Shadow

"This is none of your ga-dam business." Joe said gesturing toward the letter, "And I'll thank you not to go messing around in my affairs! When I want your help, I'll ASK FOR IT!" Joe spat these last words as he rose from his seat.

Rob stood there not saying a word. Then after a long moment he looked Joe in the eye and spoke in a calm voice.

"You're right *Captain*, my apologies. It isn't any of my business, *none* of this is... Sir." and placing his cap on his head, Rob came to attention, saluted Joe and left the pub slamming the door behind him.

Joe stood there in silence while the rest of the men in the bar stared, then resumed their conversations. Uncomfortably, Joe sat back down at the table.

Several hours and drinks passed while Joe sat alone in the pub mulling over his anger at Rob.

"The *audacity* of the guy. The *nerve*!" Joe muttered to himself, his thoughts blurring through his alcoholic haze. "Intercepting a private letter and meddling in my private affairs." He had told no one about this being his birthday.

"All this time worrying about what I'd done or thought I'd done for what? Nothing! And all the time he knew, never saying a word, letting me go on suffering that way!" Joe was furious for what Lt. Robert Browning, his wingman, his *friend* had put him through.

Downing the last of his sixth drink, Joe picked up the letter he'd written from where Rob had tossed it and tore open the envelope. Unfolding the letter, he began reading and as he did, his hands started to shake. Seeing the words he had written to Mo in the heat of anger his mouth went dry. What had he almost done and what had he now done? Joe slumped in his seat.

The night air was clammy, but it helped ease Joe's restlessness while he walked back to the base... alone. He had been wrong, BOY had he been wrong. He'd misjudged the situation and sent the wrong message again, but this time no one had intervened and the message had been delivered *loud and clear*! He needed to find his wingman and apologize.

Lighting a cigarette, his mind wandered back to the day he'd carried the pails of milk to the Fowler's farm and to that first night he and Rob had walked back from the pub after their fifth mission. That was countless missions ago and for well over half, Rob had been flying his wing, always looking out for him, protecting him and keeping him safe. Yeah, sometimes Joe resented the *way* Rob had kept an eye on him, but he had always been there when they flew, it only stood to reason that he'd do the same on the ground. Joe felt ashamed of the way he'd acted.

## Robert Brun

Back at the base, Joe walked over to Rob's hut and could see the light still on through the window, but somehow he just couldn't bring himself to go in. After standing there for the better part of a half hour, he gave in to his uncertainty and returned to his own hut. They were both scheduled to fly the next day and Rob would never skip a mission, he would talk to him then. Joe hoped he would find the right words.

Back inside his own quarters, Joe collected his thoughts and taking out pen and paper, and choosing his words very carefully, he began to write.

*My Dearest Maureen,*

*Please forgive me for not having written for far too long, but things over here have been difficult. I can't go into detail because if I did, the censors would black it out anyway, but if you've been following the news, you have an idea.*

*Mostly, I just wanted to let you know that despite everything going on, I still think of you often, miss you very much and look forward to seeing you again when this is all over.*

*With Love,*
*Your Pilot, Joe*

Joe reread the letter several more times before folding it and carefully slipping it into the envelope. Although it was late, he grabbed his coat and headed over to the base post office where, as luck would have it, the Mail Clerk was just loading the outgoing mail into the back of a truck.

"Got one more for you." Joe called to the Sergeant, handing him the envelope. "Special delivery."

"Ain't they all?" The Sergeant called down in reply.

"I guess so." Joe thought.

# My Shadow

## Chapter 31
## Flight Home

*Friday April 6, 1945: Awful day, particularly after the night I had. I'm still not certain what happened, but I'm not in the mood to find out. Maybe after today's mission, maybe - R.B.*

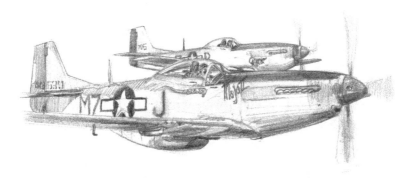

Rob awoke in a bad mood. The dressing down he'd received the night before, left him with no interest in the company of his Squadron Commander. Oh, he'd fly his wing on the mission all right, but that was it!

"That ungrateful son-of-a-bitch." Rob thought still bothered by Joe's biting words of the night before.

"It costs me six packs of smokes to get that love sick bastard's letter back from the base Mail Clerk and that's the thanks I get? To hell with him!"

Rob kept his distance all morning, skipping chow, sitting in the far corner at the mission briefing and avoiding Joe throughout the mission delay and later, on the flight line.

After forming up, Rob took his usual position as Red Two, 5,000 feet above the bombers with Joe just slightly ahead and below him.

The flight to the target had been a bore and for the first time, Rob was glad about radio silence. He had nothing to say to his Flight Leader.

Listening to the drone of the Merlin, Rob began to replay the events of the last two months in his mind. Maybe it hadn't been any of his business, but he knew something was up with Joe that day he blew by him holding that letter. Rob had followed Joe to the mailroom and couldn't help noticing the ferocity with which Joe had shoved the letter through the mail slot. After retrieving it, however, Rob found he was now in a real jam. Having read what Joe had written, what was he going to do? He couldn't send it, what would Mo say? And he couldn't return it, what would Joe say? Then an idea hit him.

## Robert Brun

If, he reasoned, he just kept the letter, Mo was sure to write again sooner or later and when Joe saw it wasn't a *Dear John*, he might be confused, but at least he would have had a chance to calm down. It seemed like a good idea at the time. It never occurred to him that Mo wouldn't write for this long, and Rob could see the effect this was having on his friend. Nor had he anticipated that Joe's reaction to his *help* would have been so negative. Why shoot the messenger, after all, he'd done the man a favor... hadn't he?

Rob's ruminations came to an abrupt end when the first enemy fighters appeared rising up from the clouds below. Ignoring the escorts, the enemy planes headed straight for the bombers.

Although having done it countless times before, Rob had never liked flying into these massive bomber formations to engage the enemy. There was always the very real possibility of collision to think about. Also, each of those heavies had ten or more 50-caliber machine guns manned by gunners shooting at any fighter within a thousand yards. He didn't blame them for being overly cautious, but it made him feel uneasy just the same. It always reminded him of running through the lawn sprinkler as a kid. "Solid lead rain drops." Rob thought following Joe into the attack.

"Tanks away. Follow me. Let's go get-em." came Joe's steady voice addressing Red flight. Radio silence now broken, a running commentary began accompanying the ensuing battle.

Rob didn't reply, but just clicked his mic. He was still upset with Joe, but now with more important things to do, his anger began to subside. He released his wing tanks and switched to internal fuel. "The *Shadow* knows."

"Red three with you."

"Red four, roger." Came Taylor and Wilson's replies.

"Cap, you think you could ask our Big Friends to watch out who they're shootin at? One of those fortress gunners nearly punched my ticket last week." Rob overheard Taylor call.

"Just keep in tight and worry about those 109s."

Taylor clicked his mic in a response that conveyed his annoyance.

Rob watched Joe drop the nose of his Mustang and dive straight into the massive swarm of German fighters. A faint stream of black smoke appeared from the exhaust stacks as Joe fire-walled the Mustang's engine through the gate, the thin wire stop that activated the Merlin's water/methanol injection system.

Adjusting his speed to stay with his Flight Leader, the two planes drove forward into the fray, Taylor and Wilson breaking left together to cover their flank.

## My Shadow

The chin turrets installed on the newer 17s and 24s had offset the head-on attack of the Luftwaffe, but these attacks were still effective. Closing on the bomber at speeds of more than 500 mph, the German pilots had only seconds to aim and shoot before having to break off and dive away or risk certain collision. Then, rolling onto their back to expose the armor-plated belly, the pilot would pull hard on the control stick and execute a positive g-force dive under the nose of the plane. The aim had been good and the lead bomber of the Circle `B' group dropped out of formation, its cockpit shot to pieces.

Following Joe down at full speed, Rob could feel his stick start to shudder while watching the air speed indicator climb past 475 mph. At 1,500 yards, Joe opened fire sending six streaks of tracers spraying down in front of the lead bomber. Knowing that he was out of range, it nevertheless had achieved the desired effect and the 109s broke off the attack peeling off to the right.

Joe threw his plane hard over to follow, and Rob stayed tight on his tail all the while keeping his eyes open for enemy fighters.

The scene was chaotic. German and American fighters; white stars and black crosses by the hundreds all flying in every direction, spinning and diving between scores of bombers holding to their steadfast course. Tracer rounds flashed everywhere.

Turning his head back around, Rob spotted Joe locked in pursuit of a Bf 109 closing on a B-17. Joe's bullets were getting hits all around the cockpit of the 109, but still the German continued to fire at the crippled Fortress. Then, with a bright flash, the 109's right wing separated from the plane and the fuselage began spinning downward out of control.

"That makes lucky number 13, Georgia." Rob thought, the two planes breaking left and right to avoid the closing bomber. The victory, however, soon lost its sweet taste. The Fortress, with two engines on fire, drop out of formation while one, then two more chutes appeared. Rob was too busy to wait for the seven more he hoped would come.

Circling, Rob glanced over to see *MoJo II* banking sharply to catch up with the rest of the bombers beginning their runs over the target.

"You there, Rob? Over." Joe's voice came over his headphones.

"Got your six o'clock, Georgia." Came Rob's stock reply.

The two fighters reformed and assessed the situation. Red three and four, Taylor and Wilson, soon followed. The enemy fighters were gone... for the moment.

Over the target, the flak was heavy. The German fighters had left the area allowing the anti-aircraft gunners on the ground to take over their job. Red flight regrouped with the rest of the squadron above and off to one side watching while two more bombers, hit from below, began to fall. A feeling of impotence washed over Rob while observing the carnage taking place beside them, but there was nothing he could do but wait and see who remained for the trip back.

Holding their course straight through the bomb run in order to achieve maximum bombing effect, the Fortresses gave the German gunners perfect targets and the 88 and 105mm anti-aircraft batteries. By the time Rob and Joe reformed with the bomb group, 16 more planes, 160 airmen had gone down.

"Rob, I'm low on ammo. Heading back up to keep an eye on the Big Friends. They look like they could use a bit of company. Over."

"Roger, Red Leader. I've got a few rounds left, but I'd better hang onto them in case you attract any more flies."

"We'll head over to check the tail markings on those 17s, give em a safe ride home. I'll call the Big Friends."

Switching over to C-channel, Rob listened in while Joe contacted the lead bomber group.

"Little Friends to Big Friends, are you happy? Over."

"Roger, Little Friends, a bit beat-up, but on time and on course."

"Roger Big Friends, we'll hang tight and keep our eyes open. Mad-dog Red Leader out."

Switching back to the fighter channel, Joe called to the rest of squadron.

"Phoenix Squadron, this is Mad-dog Red Leader." Joe radioed using that day's squadron's call sign. "Stay high and tight with the Big Friends. I'll take the rest of Red Flight and keep an eye out for bogie's."

Still over German territory, the Mustangs had just begun the long trip back to the coast when the second wave of fighters appeared. Hundreds of them, refueled, rearmed and mad as hell!

They seemed to come from everywhere and all at the same time. Before Rob knew what had happened, he saw four FW-190s closing fast from above and behind.

"Red flight, break!" Rob shouted.

While watching the fighter's approaching from ahead, Rob had missed the group coming in behind, out of the sun. A classic out-flanking maneuver and he had missed it.

The suddenness of the attack did not allow Rob time to chastise himself; there'd be time for that later. He circled around with the three other planes, back into the fight.

Joe attacked an FW-190 from below, while Rob positioned himself behind and to the right, pulling around and approaching from an acute angle that gave the best vantage point to keep an eye, not only on his Flight Leader, but also for any bandits that might get on Joe's tail.

## My Shadow

Before Joe had a chance to get off more than a few rounds, the 190s broke for the deck and headed for home.

"Live to fight another day? Let 'em go, I'm out of ammo anyway. Over." Joe called to his wingman, the two Focke-Wulfs disappearing below. Red One and Two climbed back up, rejoining the bombers.

There wasn't much time to think before the rest of the German *Schwarms* dashed in like swallows through the formation, but despite their superior number, most had Mustangs in pursuit, and many were already trailing smoke.

"Mad-dog Leader to Phoenix Squadron, look's like we have them on the run, but don't stray too far from the Big Friends." Then came Taylor's panicked voice.

"He's on me!!! I can't shake him! He's on me!!"

Twisting his head all around, Rob spotted Taylor's Mustang below and off to the left. It was smoking badly and losing altitude fast followed closely by the 109 that had put him in that state, his wingman, Wilson was nowhere in sight.

Rob keyed his mic even before the words had formed in his mind, "Georgia... Taylor... he's..."

"GO!" was all Joe had to say and Rob dropped down in pursuit of the two diving aircraft.

Pushing his plane over and accelerating the Mustang to full war emergency power, Rob listened to the increasing whine of the engine as he headed for the deck closing with the pursuing 109 with each second. As if in slow motion, Rob watched the 20 mm cannon fire from the 109 shredding Taylor's left wing and elevator, smoke belching out from beneath the engine cowling.

Spent German shell casings were striking his prop when Rob at last opened fire giving the 109 a few short bursts, but the 109 pilot fixated on his target, took no notice. Rob fired again, but to the same effect; the German completely ignored him.

Sweat began to form on Rob's forehead as flames began licking off Taylor's plane. Time was running out and Rob knew he had to act fast. He had no other choice.

"Break off, you *Brat* Bastard, break off dammit!" Rob shouted in exasperation then gripping the control stick, pulled hard on the trigger and gave the 109 an extended burst.

Tracers came in rapid succession indicating Rob was expending the last of his ammunition as he raked the 109 from engine to cockpit. The German pilot had ignored his tail and the mistake had cost him; his plane was now going down in a fiery ball of smoke and flames.

## Robert Brun

Pulling his fighter quickly to the right, Rob skirted the flaming Messerschmitt and scanned the sky for Taylor's Mustang. To his horror, he watched the plane slam straight into the ground and explode.

Rob stared at the wreckage when something caught his eye. Silhouetted against the rising black cloud of smoke, the white canopy of Taylor's parachute rocked slowly from side to side while descending to the field below. Riding the `silk elevator,' Taylor made it safely to the ground.

Relieved, Rob circled the downed pilot banking low and saluted Taylor who, collecting his chute, returned the gesture.

Heading back to the battle in a slow climbing turn, Rob saw the remains of the 109 crashed in a nearby field. Somehow the German had managed to land the burning plane and crawl from the wreckage, but with his body completely alight, made it only a few yards before succumbing to the flames. Witnessing this gruesome sight, Rob's whole body started to shake.

"You stupid *Brat* BASTARD!" Rob shouted. "Why didn't you break off?" And staring at the remains of the burning pilot, a sudden realization hit him like a bolt of lightning; he had <u>wanted</u> to kill this man!

Rob's mind reeled. The *Shadow* had killed! Not a plane, not a truck or a train, but another human being. Not with random bombs dropped on unfortunate combatants, but a hated enemy within his sight. His delusional game was over and in an instant Rob had gone from protector to Judge, Jury and *Executioner!* The desire he felt to kill this

## My Shadow

man terrified him. He had wanted him dead. Dead for Taylor, dead for Cushman, dead for Harrison, for Adams and... And for...Paula! What had he become? A killer, a monster, just like all the others, just like... Joe!

Rob was suddenly filled with self-loathing. The hideous image of the pilot's burning body flooded his mind. Someone's son... father... brother... friend, lover was dead because of him and he'd seen it with his own eyes. Wanted it!

The image of Paula's motionless body appeared from deep within his mind and her eyes opened, pleading with him and asking why? Why hadn't he found another way?

Waves of nausea gripped Rob while he scanned a sky that spun wildly before him, searching for his flight leader, the rest of the squadron; anything his mind could latch onto.

Rapid breaths escaped his lungs and feeling suffocated, Rob ripped the oxygen mask from his face while rivulets of sweat poured down into his eyes. Through blurred vision he forced himself to focus on the instruments trying to gain any emotional foothold. Then, flying straight and level, his mind began to slowly clear. He replaced the oxygen mask and began taking deep regular breaths and heard Joe's voice in his headphones.

"Mad-dog Red leader to Red two, are you there Rob?"

With a shaky hand, Rob keyed his mic.

"Mad-dog leader, this is Red t..."

## *BANG!*

The loud sound reverberated through the fighter from behind the armor-plated seatback and the radio went dead. The altimeter was the next to go and one by one the gauges of the instrument panel shattered. The throttle quadrant exploded in his hand and Rob felt something hard slam into his left thigh. The sensation was unreal, the pain intense!

Rob dropped his right wing and skidded the Mustang to one side. In an instant, the first FW-190 flashed by while he fought to regain control of the spinning fighter. Before he had a chance to think, a second round of shells raked the fuselage, shattering the rear of the canopy and blasting him with frigid air.

"Let's get the **HELL** Outta-here" Rob shouted fighting off the agony that shot through his left leg and pulled his spinning plane around.

Scared and yet focused, his heart pounded and red flashes clouded his vision while he fought to regain control of the tumbling fighter. Outside the cockpit, the world spun and Rob turned to his shattered instruments for guidance. When *Shadow II* recovered from the spin and he looked up again, the world had gone grey; the pounding the plane was taking had stopped.

Robs first thought was he was dead; everything had become so peaceful, but the frigid wind blasting from behind indicated otherwise. He had fallen into the clouds, and was safe... for the moment.

Trying to steady both the plane and his nerves, Rob collected his thoughts. Although he could barely see beyond the cowling of his plane, it also meant he was invisible to the Focke-Wulfs. Making a few adjustments to the control column, he leveled the plane the best he could without the benefit of the artificial horizon, which had shattered along with the rest of the instrument panel.

Assessing the situation, Rob called upon all his training and experience to keep airborne. Then *Shadow's* engine started to whine and the airspeed indicator, one of the few gauges still working, began climbing. Confused, Rob maneuvered the stick in an attempt to regain control, unsure what was happening?

As soon as the question entered his mind, *Shadow's* cockpit lit up with such intensity that Rob was certain he'd caught fire. His sudden panic passed when he discovered that, experiencing vertigo, he had lost his sense of which way was up and exited the clouds heading straight down!

Fighting both gravity and pain, Rob wrestled with the flight controls managing to level his plane once again. The success, however, was short lived for no sooner had he regained control when black puffs of anti-aircraft fire began appearing all around him. His exit from the cloudbank had brought him out directly over a German airfield and the base defenses were now targeting him.

Unable to think of anything else to do, he pulled back on the control stick and returned to the safety of the clouds. The Mustang complained bitterly as the world around him once again became an opaque curtain of grey. He was again completely blind, but at least the *ack-ack* had stopped.

Rob was now fighting desperately to remain conscious while struggling to keep his damaged fighter in the air. Oil streaked the canopy and the engine sputtered while he fought to keep the shot up Mustang flying, mostly by force of will. The smashed instrument panel made it impossible to be certain if the oil gauge was correct in its reading of zero pressure, but he guessed it wasn't far off. The 190s had hit him hard and he cursed himself for having not seen the FWs sooner.

"Some wingman?" Rob thought aggravated with his conduct. He'd really dropped the ball this time... Letting his anger get in the way of his responsibilities, he'd lowered his guard. Now he was limping back to England in a plane that, for all his training and experience, shouldn't even be flying. There was a pulsating throb in his left thigh from the explosive

## My Shadow

shell fragment that had taken out the throttle housing and a second bullet had torn into his foot. He could feel a warm, sticky wetness in his boot that he knew wasn't sweat. Just trying to keep pressure on the rudder pedals sent a sharp, burning pain up his leg, but that pain was now the only thing keeping him conscious.

With some difficulty, he managed to secure the headphones wire around his leg, stemming the flow of blood from the wound that soaked his flight suit. The radio had been the first casualty of the attack, so it no longer mattered that the microphone cord was now being used for something less than regulation.

Flying on, Rob considered his limited options. He could head for neutral Switzerland where, if he made it over the Alps and managed to land in one piece, meant arrest and interment for the duration of the war.

He could ditch in the North Sea, but the thought of betting his life on Air-Sea-Rescue's 20% success rate for finding downed pilots, didn't appeal to him either.

Crash landing behind enemy lines was another possibility, however, with his leg in its current state, he wouldn't be able to walk much less run, and he had no intension of hand delivering his Mustang to the *Wehrmacht* even if it was a bit banged up. Rob increased the oxygen flow and cleared his head; at least that system was still working.

For the next hour, Rob dodged in and out of cloudbanks hiding from enemy planes and ground fire, all the while fighting to remain conscious and continuing the monumental task of keeping the damaged *Shadow II* in the air. The shattered canopy offered little protection as the frigid air numbed him to the bones. The overcast finally began to breakup when the Dutch coastline came into view. Just what had kept him going this long, Rob didn't know?

The western sky had turned from blue to orange by the time Rob glanced at the cockpit clock making note of the time; 20:23, he'd been in the air for over six hours, but without an operating fuel gauge, he could only guess how much gas he had left. Rob leaned out the fuel mixture even more. This would overheat the Merlin, but with the state of the rest of the Mustang, it no longer mattered and the shattered plane droned on.

Thinking back over the mission, Rob was again struck by the image of the man he had killed. Dead, a man had died a horrible death and he had caused it... wanted it. The memory of the hatred he'd felt sickened him, but what choice did he have? Taylor would most certainly have been killed if he hadn't acted. Was it even sane to have let Taylor, his squadron mate and friend, be killed during a war just to keep a crazy promise he'd made to himself? And how would he now feel if it were Taylor and not the German pilot who was dead? He had no good answer.

# Robert Brun

Rob didn't have time to ponder these questions long, when the engine began to sputter and cough.

"This is it!" Rob thought. "Time to hit the silk."

Rob fought pain trying to recall his training.

The theory for safely exiting a fighter was to trim the plane to full nose-down then kick the control column hard forward as you jumped. This allowed the downward trim and reverse lift created by this maneuver to throw you clear of the tail section. He'd heard about pilots that had been injured or even killed hitting the rudder at 150+ miles per hour and with his injured leg, what were his chances?

The Merlin continued to cough and now at 5,000 feet, Rob removed his oxygen mask. Holding the control stick of the untrimmed plane between his knees, he unfastened his safety harness and slid what remained of the canopy open, then mustering all of his will and trying to ignore the searing pain, Rob prepared to jump.

The plane wobbled slightly when he placed his good leg against the smashed instrument panel then looked down over the landscape of occupied Holland when a strange thought occurred to him.

"They don't like me down there." he thought, stating the obvious.

The idea of spending the rest of the war in a German POW camp held no appeal for him. The stories and rumors of pilot hacked to death by enraged villagers caused an involuntary shutter and the briefings about what to expect if captured all ran through his mind. Rob wanted no part of either, but mostly, he needed to know that Joe was all right. He was the *Shadow*, *MoJo's* wingman, and like it or not, regardless of last night, Joe was still his responsibility.

"Dammit!" he said aloud and made up his mind right then to make it back to Abington... Somehow, or die trying! With a great deal of pain and difficulty, Rob managed to get himself buckled back into the plane.

Trying not to think about the wide stretch of the frigid North Sea that still lay ahead of him, he re-trimmed the Mustang and took the control stick in his knees. Using both hands, he lifted what had once been the engine throttle controls while the Merlin continued to cough.

"What a mess!" Rob thought while sorting through the broken pieces trying to determine which operated what. Grabbing the largest cable of the bundle, Rob pulled hard on it and drew a deep breath when the coughing Merlin slowed to just above idle speed. Realizing his mistake, Rob snatched another cable of the same size and gave it a sharp yank, the engine speeding back up again to its irregular sputtering.

Another tug with his frozen fingers and the Merlin picked up rpm's and seemed to smooth out just a bit. Heaving a sigh, Rob wiped the sweat from his face and looked down at his glove to see it was covered with blood.

## My Shadow

"Crimminy!" Rob thought out loud, "What a mess! Is Tom ever gonna be sore at me."

Turning the fighter slightly, Rob was relieved to note that the flight controls, although mushy, were still holding together and responsive. The compass too, by some miracle, was also operating. Rob pointed the nose of the plane due west and readjusting the fuel flow as lean as he dared, flew *Shadow II* out over the North Sea toward England on what he knew was going to be a long slog home.

# Robert Brun

## Chapter 32
## Arrival
*Friday, April 6, 1945*

    Ignoring the cold, Joe continued to stare into the darkening sky. Rob was already three hours overdue. Where was he? The last he'd seen of his wingman, he had two 190s on his tail. Joe had tried to intervene, but had lost them in the clouds.
    Joe thought back to the night before and winced. He'd got drunk and lost his temper, but why? For as long as he'd known him, Rob always seemed to appear from out of nowhere, so why had his interference angered him so much this time? But Joe already knew the answer. He was the Squadron Commander; he was in charge and led the pilots during missions, he didn't need someone looking after him... or did he?
    Up in the sky, the men were his responsibility, but on the ground things had been different. His inability to deal with his personal problems had made him feel weak and Joe resented Rob for seeing it; it was his job to lead and he should be able to handle his own problems. But Rob had seen though this facade and tried to help. The hard truth was, he had. Rob always protected his Flight Leader in combat and this time he'd protected him from himself.
    Joe now understood what Rob knew all along; every Flight Leader needs his wingman. The realization made Joe feel petty. Maybe he wasn't as *grown-up* as he wanted to believe. All the *should-haves* came flooding back.
    Angry with himself, Joe continued to stand, fists clenched tight while behind him, a staff car pulled up and Colonel Tomlinson got out.
    "Captain?" the Colonel said softly but sternly "Any news?"
    Joe turned and addressed the Colonel. "Nothing yet, but I'd like to wait a bit longer."
    "Captain. The Colonel spoke with a sigh. "I know you're worried about Lt. Browning, but there's not a damn thing to be accomplished standing here all night. You've got other men in the squadron depending on you. You're leading the mission again tomorrow and I need you at your best."
    Joe didn't respond, but the Colonel's words bit deep into him. He knew his friendship with Rob had affected his judgment, but somehow leaving the field felt too much like abandoning his wingman and that was something he wouldn't do. Rob had never abandoned him even when Joe had tried to drive him away and he wouldn't abandon Rob now.
    "With all due respect Colonel, I'd like to wait a bit longer, Sir."

## My Shadow

"Captain, you've lost men before and there's nothing you or I can do about it, both life and this damn war goes on, this is no different. We do our jobs and when it's over, we go home."

"You talk as though Browning is already dead Sir!" Joe spat staring the Colonel right in the eye.

"Until this war is over Captain, we're all dead!"

There was a long moment of silence. Colonel Tomlinson was his Commanding Officer and had earned Joe's respect, but Rob was his friend and somehow that counted for much more.

"Colonel I..." Joe protested, but stopped himself before speaking his mind. Right now all that mattered to him was the whereabouts of his wingman. There was another awkward silence and Joe wondered just how far he could push the Colonel before he crossed the line of insubordination.

Then he heard it, softly at first, but with each second the sound grew louder and its source became unmistakable. The sputtering and coughing could not disguise the sound of the Merlin engine fighting its way back to the airfield. The two men turned, staring at the small, distinctive silhouette against the darkening sky. The plane grew larger above the tree line and Joe could see a plume of smoke trailing behind the plane. Without a moment's hesitation, Joe knew what had to be done and sprang into action.

"Quick, get the lights of that vehicle on!" he barked pointing to the Colonel's staff car and forgetting to whom he was speaking. Tomlinson reddened at the tone of the Captain's order, before realizing that Joe had something he didn't, an idea, one that might save this pilot's life.

Grabbing a tire iron from inside the jeep, Joe ran around to the front of the vehicle and pried the blackout covers off the headlights. Then running over to the Colonel's car, he did the same thing. The area in which the men stood was now brightly illuminated.

"Take the Colonel's car over to the other side to the field across from mine." Joe instructed the Colonel's driver over the growing sound of the Mustang's failing engine, and jumping into the jeep shouted.

"Set up the headlights so the beams intersect in the middle of the runway... and hurry!"

Now, almost frantic, Joe stomped on the accelerator pedal and spinning the tires in a spray of turf, left the Colonel standing alone as he took off down the left side of the runway to mid-field.

Turning the wheel hard to the right, Joe felt two tires leave the sod briefly as the jeep skidded to a halt. Looking across the landing strip, he was relieved to see that the Colonel's driver had understood his instructions and duplicated his actions. Both vehicles in position, the headlights now formed an "X" of light at the center of the runway that he hoped Rob would see and understand.

Grasping the steering wheel so tightly his knuckles turned white, Joe saw the shape of the Mustang clearly now as it descended, beginning it's final approach.

The moon had just crested the tree line and the word *"Shadow II"* shown along the side of the cowling.

"Com-on buddy" Joe thought out loud. "You gotta make it!"

It was difficult to assess the damage to the aircraft floating down toward the end of the field, but through the belching smoke, Joe could see the flaps had been lowered, but only one of the two main landing gear was down, the other sticking out only part way. Despite strong feelings to the contrary, Joe knew that he had done everything that he could, the rest was up to Rob. All Joe could do now was watch... hope... and pray!

The crippled Mustang sputtered its way down to treetop level descending rapidly, its one extended landing gear barely clearing the perimeter fence at the far end of the field.

Like the sound of a dying animal's last breath, the failing Merlin gave one last roar of power lifting *Shadow's* nose five degrees and pulling the plane skyward one final time before sputtering to a stop. A cold silence descended over the field.

Only a slight rush of air was audible as the damaged fighter dropped its right wing, side slipped and touched down on the one good main wheel right in the center of the crossed beams of light.

When the tail wheel made contact with the field, the torque produced by the angle of the landing proved too much for the weakened airframe. Bending the main wheel strut under, the left wingtip dug a deep furrow along the grass field putting the plane into a slow counter-clockwise spin.

## My Shadow

The sound of tearing metal was sickening as the plane's fuselage twisted and the aluminum skin of the left wing tore and broke free. The plane then came to an abrupt stop, and everything was silent.

Joe sat frozen for only a second, the wail of the fire rescue trucks breaking the silence. Without thinking, he sprinted across the runway to where the smoking plane now lay broken and still. Even in the faint glow of the rising moon, Joe could distinguish between the battle and crash damage of Rob's plane. A strong aroma of glycol coolant from the ruptured radiator filled the air when Joe approached the plane, then he stopped in his tracks. He had never seen a Mustang shot up as badly as the one that lay before him.

*Shadow II* was riddled with bullet holes and flak damage from nose to tail. Long streaks of oil ran along the side of the engine cowling covering the windshield. Several large holes from explosive cannon shells dotted the fuselage and wings. Half the lower leading edge of the rudder and elevator had been completely shot away. The entire tail section of the plane, from the radiator on back, had broken from the force of the landing and bent off at a hideous angle. The radio compartment behind the cockpit was a shambles.

The soft tick of cooling metal snapped Joe out of his daze and he approached the plane unsure if he wanted the answer to the question that now haunted him. Sirens wailing, Joe took a deep breath and climbed up on what remained of the port wing.

Peering through the shattered Perspex canopy, the inside of the cockpit was dark, but for the dim light from the instrument lamps that cast an eerie glow on the seated figure inside. Joe reached in across the cockpit, grabbed the emergency canopy release lever and pulled. What remained of the broken canopy came free and Joe pushed it off to the side relieved to feel it fall away.

Inside the cockpit, he could make out the shape of Rob's body slumped over in the safety harness. Trying not to give in to his worse fears, Joe reached out his hand and felt Rob's shoulders move. Raising his head with a groan, Rob looked up at Joe.

"Hey there, Georgia," Rob said softly through bloodied lips. "Not one of my better landings, huh?"

Joe involuntarily gasped at the sight of his wingman's face. Rob's head had struck the gun sight on landing and fresh blood seeped from a deep gash across his forehead. His nose appeared to be broken, the swelling rapidly distorting his features.

"You stupid son-of-a-bitch!" Joe blurted at the relief of seeing his wingman alive. "Do you have any idea how worried I've been?"

"Yeah.., I guess I do." were Rob's only words before passing out.

# Robert Brun

## Chapter 33
## Boys to Men
*Monday, April 9, 1945*

For the next two days, Rob lay unconscious in the base medical ward. The blow to the head had resulted in a subdural hematoma and Rob remained too critical to be moved. The piece of shrapnel had been removed from the left upper thigh and a 7.92 mm. bullet from his foot, but it was the coma that was most worrying. Doc Lewis assured Joe that he was doing everything he could for the man, but that he was in pretty bad shape. It would be touch-and-go for a while.

Rob had also lost a lot of blood and Joe's concern turned to frustration when he learned that he was not a compatible blood donor. A bottle of scotch and his week's ration of smokes solved the problem by obtaining the squadron's medical records from an orderly and successfully persuaded everyone on the base with a matching blood type to donate. All complied and none complained.

The following day, Joe was flying once again. He had been assigned Lt. Jefferson, the enthusiastic new arrival from breakfast the previous morning, flying as his wingman. After the briefing, Joe had been uncharacteristically harsh with the new pilot's inquiries while approaching the flight line. Disgusted with himself for the way he had acted, Joe knew full well that Jefferson had only come over to investigate his Flight Leader before his first mission, but Joe had torn into the young Lieutenant for some minor infraction. Joe knew exactly why though, Lt. Jefferson was not Browning.

Once airborne, things weren't much better. Joe was uncomfortable at not having Lt. Robert Browning flying his wing for the first time in almost a year. He had become accustomed to seeing the *Shadow* in his peripheral vision when he flew. Maybe he had been a bit hard on this rookie, earlier, after all he was just trying to make his *first five* and Jefferson still had a lot to learn, as they all once had.

## My Shadow

As April wore on it became clear from the lack of activity in the skies over Germany that the *Luftwaffe* was finished, but like the Nazis themselves, although beaten, they continued to fight on. The endless pounding German industry was receiving had reduced the supply of oil and aircraft fuel to a trickle. It was also clear from German combat tactics the pilots they sent into battle had not received adequate training. Joe remained vigilant however, even a rookie pilot could get off a lucky shot and no one wanted to *buy the farm* this late in the game.

Even the Me-262 jets that occasionally appeared, once you got over the initial shock at their speed, posed no real threat to the massive allied bombing formations being sent over almost daily.

Sitting in his Mustang at 30,000 feet, flying lazy S-turns above a seemingly endless stream of bombers with nothing more than 1/16" of aluminum between himself and 40 degrees below zero, Joe sat trying to stay warm and alert.

Radio silence was still the order and the drone of the Merlin, which he used to find so reassuring, now seemed to be mind numbing and Joe found he had to fight just to keep awake.

Down below, the bombing, that had been going on day and night for almost a year, had reduced the once beautiful German cities to smoldering rubble, visible even from this altitude. The Russian Army had crossed the Oder River and was now at the outskirts of Berlin, but still the Germans fought on. Now, even tired of flying, Joe thought of Mo back home in the States, building widgets for the *Arsenal of Democracy* and wondered again who and what he would find when he got back.

Anxious about the approaching end to the conflict and his return home, Joe worried how Mo had changed over the last two years, and how had he? Would they still *fit* as they once had? He had seen and lived through so many things that he didn't even want to remember much less talk about. And what about the men he had personally killed?

Joe had seen chutes coming from some of the 15 planes he'd shot down and as odd as it seemed, it always came as a surprise when the canopy flew off and a human being jumped from these enemy machines, but what of the times when there had been no chute? In the heat of battle, it wasn't always possible to witness the final results of his actions.

Joe didn't want to think about it. Just the same, he knew he'd be deluding himself if he thought they'd all floated down safely to earth. Many of the 597th's men hadn't made it back and in this respect the German pilots were no different. He had a job to do and so had they. He had survived they hadn't been so lucky.

## Robert Brun

Somewhere deep down inside, Joe knew he would be able to live with what he had done, after all, he was a fighter pilot and this was a war, but right at this moment, he couldn't imagine how.

Looking down at the formation of bombers, Joe wiped the frost off the inside of the canopy with the back of his glove and watched. Once again as the bombs fell on the city below, their noiseless explosions covering the ground. He watched too as black puffs of exploding anti-aircraft shells dotted the sky within the bomber formations. The white-hot, razor-sharp metal that ripped apart both aircraft and crew and had taken Captain Adams on that first mission. Joe was selfishly glad to be above it all. The thought made him angry and he closed his eyes and looked away.

On the return trip, Joe again thought about Rob... About his wingman and friend lying in the base hospital unconscious, touch and go! He thought about all the things they had been through together, good and bad and how Rob had always seemed to be there when he had needed him and gone when he didn't. *The Shadow,* he laughed, maybe there was more to that then he thought.

"Mad-dog Red Leader?" came a hesitant voice over Joe's headphones. " This is Red two. I think I see something. Over."

It was his wingman, Lt. Jefferson breaking radio silence.

"Let's have proper radio protocol Lieutenant? Over." Joe responded, annoyed at being disturbed from his melancholia.

"Uh, I think I see, Um... Mad-dog Red Leader, I've got two, no three bogies, ten o'clock low. Over."

Joe swiveled his head around and spotted three small dots off to his left and below the bombers; there was no mistaking the black crosses, these were enemy fighters!

"Well I'll be damned." Joe thought feeling a bit embarrassed. "Lieutenant Jefferson, you got eyes after all."

"Good job, Red two, time to go to work. Red three, this is Mad-dog leader; Red two and I are going down to engage bandits. You and Red four stay with the Big Friends. We'll call you if we need help. Red two, tanks away!"

Joe switch to internal fuel, pulling the salvo release levers that jettisoned his wing tanks and swung the Mustang up and around to engage the enemy planes.

"Stick close to my wing, Red two and keep your eyes peeled for anymore company. Over."

Maneuvering his plane to put the sun at their backs, Joe glanced quickly into the mirror impressed to see that Jefferson was already in position. While the two Mustangs approached, Joe noticed that the three 109s made no change in their course or speed.

## My Shadow

Lowering the nose of his fighter and lining up his sight, Joe depressed the trigger and opened fire at 100 yards. The rearmost plane, beginning to smoke, rolled over and spun earthward out of control, while the two remaining 109s broke wide, Joe and his wingman flying between them. Pulling back hard, Joe felt the bladders of his G-suit compress his leg muscles and watched the more distant of the two 109s dive sharply and break for home.

"Let him go Red two." Joe instructed his wingman while falling in behind the remaining Messerschmitt, but before he had time for a second thought, the 109 pilot had chopped his throttle. Joe snap rolled to avoid a collision and recovering control of his plane, maneuvered to reacquire his target.

The 109 was turning sharply when Joe pulled in behind to join the *Lufbery*. This classic German maneuver of getting into a turning duel was an early air combat tactic developed during the First World War and used again during the Battle of Britain.

Each plane friend and foe flies in a circle behind the other in order to cover the others tail. Because of the tight angle of bank, none of the planes can get into a firing position. Any plane, however, that breaks the circle can easily be followed by his opponent.

Although, by now somewhat outdated, it was still a fairly reliable tactic just the same. This 109 pilot had, however, forgotten one very important aspect of the *Lufbery*; it wasn't very effective one-on-one at low speed. This guy, Joe realized, was definitely green.

Joe dropped into position behind the 109, now flying just above a stall and could see the Messerschmitt's wing slats and flaps extended. Despite the poor choice of tactics on the part of this German pilot, Joe was unable to get into a firing position while they continued to circle.

The Bf-109, designed by Willy Messerschmitt in the 1930s and this *109-K* model, though continually modified and upgraded throughout the war, had, by April of 1945, become obsolete. This didn't change the fact, however, that it could still out-turn the P-51 at low speed and that was just what it was doing now. With combat flaps extended, elevator at full trim and throttle back as far as he dared, Joe was still unable to maneuver his plane behind the slow turning 109.

The two planes continued to circle in a descending spiral now dropping to 6,000 feet. Trying another approach, Joe eased his plane out and increased throttle just enough to pull up along side the German fighter.

Looking across as the two planes now flew wingtip to wingtip, Joe could clearly see the face of the 109's pilot. The German had removed his goggles and oxygen mask and was panting hard, his eyes wide and his head turning rapidly between his instruments and Joe's fixed gaze.

## Robert Brun

Joe was surprised to see, not the battle-hardened features of an enemy fighter pilot, but the cherubic face of a child of not more than sixteen years of age, white with fear! His powder-blue flight uniform looked brand new and spotless except for a dark patch around the collar and... The pilot was crying!

Joe hesitated only a moment before making his decision.

"No... Not today... Today, I will not kill!"

Having made up his mind, Joe gave a sharp salute to the pilot and eased the Mustang out and away from the circling Messerschmitt.

Joe had no sooner pulled away, when the wing of the 109 erupted in flames and a second P-51 raced passed.

"I got him... I got him!" came Lt. Jefferson's excited shouts over Joe's radio.

Having completely forgotten about his wingman, Joe watched the flaming Messerschmitt spin into the ground and explode.

He saw no chute.

# My Shadow

## Chapter 34
## All Right
*Thursday, May 3, 1945*

Because of the extent of his injuries, and once stabilized, Doc Lewis had Rob's unconscious body transferred to a hospital in London making Joe's daily visits to check on him impossible. Never the less, Joe made a pest of himself putting in requests for leave every chance he got.

When word came that on April 30th the *Führer* had committed suicide, it was as though the fight had finally gone out of the *Luftwaffe* altogether. The fighter escort missions, if you wanted to call them that, became long, cold, boring flights of shear monotony. Although Joe had been flying the route for over a year, now without even the chance of engaging the enemy, the missions had became little more than a chore and an uncomfortable one at that.

Returning from an early morning reconnaissance flight, Zeke informed him he had orders to report to the Colonel.

When shown into the office, Joe stood at attention while Colonel Tomlinson sat behind his desk not looking up from his work.

"At ease, Captain and have a seat." the Colonel said shuffling through some papers.

Joe took the chair and waited uneasily in anticipation of what was to come next. Nothing had been said since the night of Rob's return.

Speaking the way he had to a superior officer, behaving in the manner he did, not to mention violating blackout restrictions and deliberately damaging government property were all court martial offenses. Joe knew he had been out of line, hell, he'd known it that night, but behind it all, he still felt he had done the right thing, the only thing he could have done, whatever the consequences now.

"Captain," the Colonel started. "Do you have any idea why I asked you in here?" staring over the top of his glasses and straight at Joe.

Joe had learned since his enlistment in the military, never to offer unsolicited information to a superior officer, but this was Colonel Tomlinson, a man Joe had served with for almost two years. Somehow that made things different.

"About that night..." Joe started and before he could say another word, the Colonel cut him off.

"Listen Captain," the Colonel said maintaining his stare. "You needn't explain. Have you forgotten? I was there, I witnessed the whole thing, and despite my irritation at not having thought of it myself, I agree with your actions completely." Then he added slowly, "Under the circumstances." hardening his glare.

"I understand, Sir," said Joe feeling uncomfortable.

## Robert Brun

"...And that is why I've put in for a Commendation for your actions that night, for valor in saving the life of a fellow officer." and looking up, he added. "Not to mention a damn good pilot."

Joe sat up feeling his anxiety begin to lessen.

"And another thing, Captain." the Colonel added.

"Lt. Browning has been promoted to the rank of Captain. It won't be official for a few more days, but that's just paperwork." Tomlinson gave a dismissive wave. "And I thought you might like to tell him yourself." He added handing Joe a three-day pass. "You can catch the train to London this afternoon."

Joe was speechless. He sat there staring at the pass held in the Colonel's out-stretched hand. No court martial, no official reprimand, not even a bawling out? Joe felt like he'd recovered from a flat spin ten feet above the ground.

"Uh...Thank you Sir?" was all he could think of to say. Joe stood, took the pass and saluted preparing to leave.

"Captain. One other thing." came the Colonel's voice again.

"Ye-yessir?" said Joe stopping in mid turn.

"Give my regards to _Captain_ Swiss Cheese will you?"

"Yes Sir! I'll be sure to do that."

Joe returned to his quarters to find a letter from Mo on his bunk. Grabbing a cup of coffee from the mess, he found a sunny spot out by the airfield and sat down. Several of the men were playing baseball with a cricket bat they'd found while the occasional Mustangs came and went overhead. Opening the letter, Joe read:

> My Dearest Joseph,
> 
> I was _so_ happy to receive your letter. It had been such a long time since I'd heard from you and I've tried not to imagine the worst. I did sense from your letter that things have been tough for you also. I am so sorry that I haven't written earlier.
> 
> From what I have seen of the newsreels, it looks like things are starting to wind down over there. I do hope so.
> 
> I know they have here because I've been let go from the foundry as they scale back production. Under the circumstances, however, I can't say I'm disappointed. Working as shop Foreman has turned out not to be all it's cracked up to be and it is nice to be able to sleep in again for a change.
> 
> I hope you can see clear to forgive me if my lack of correspondence has caused you any worry. It would seem I've been mistaken about a great many things... and people... But not

## My Shadow

*about you. Through it all I have realized just what you've come to mean to me and to understand my feelings for you.*
*I hope this letter finds you well and I look forward and will be waiting for you upon your return. Hurry home Joe.*
*Much Love,*
*Mo*

Joe lowered the letter and let out a sigh. He couldn't believe his luck. Everything that had been falling apart now seemed to be falling into place. Mo knew what she wanted and so did he, now they would have the chance to find out, together. Refolding the letter, he slipped it into his jacket.

"Hey, Captain!" Lt. Jefferson called over to Joe. "You want to join the game, we could sure use a shortstop."

Joe looked up and smiled at the Lieutenant. So much of what he saw standing there reminded him of the young man he had been just two short years ago.

"No thanks, Lieutenant, I gotta catch a train."

Climbing down onto the train platform, Joe was reminded of that first morning he and Rob had arrived in London, in what now seemed like a different lifetime. God, he'd been so naive then.

The atmosphere around the station appeared much more staid than his last two visits, almost a *normal*, every-day feeling. Walking out into the bright sunshine, Joe put on his Ray-Bans and enjoyed the warm feeling on his face. Unlike that first morning, so long and so many missions ago, the sun was up and the sky was a bright blue. In addition to the people and traffic, he was surprised to hear birds singing and the sound of laughing children playing in a nearby park.

Looking down at the sidewalk, Joe could see the dark shape of his shadow cast there before him.

*"Me...., and my shaaa-dow."* Joe sang under his breath thinking back on his and Rob's first mission together. *Strollin down the av-en-nue*

*"Ay Mate?"* a passing man inquired looking at Joe quizzically.

"Oh, nothin... Just thinking about a buddy of mine."

*"Buddy..? Yeah... nyice!"* and the man continued on his way.

Joe's gaze followed the receding figure walking away with a cane in his hand and a pronounced limp. He wondered what part this man had played in all of this? How had he served? Had he been injured or wounded? Did he help put out the fires during the blitz or shoot down Heinkels over Sussex in his Hurricane? And what friends and loved ones had he lost through it all?

# Robert Brun

As horrible as it had been, Joe now understood just how they had all come to be part of this *World War!* As a result, they were now all permanently connected in a way that no one, not involved, would ever be or fully understand. How could they?

Hailing a cab, Joe climbed into the back.

*"Where to mate?"* the driver asked without turning around.

"West End Military Hospital." Joe said. "Got a buddy of mine recuperating there." fingering Rob's Captain's bars in the pocket of his tunic.

*"Gonna be al-right, will-e now?"* the driver asked glancing at Joe in the mirror and sounding concerned.

"Yep. It's all gonna be all right."

# My Shadow

## Chapter 36
## The Shadow
*Thursday, May 3, 1945*

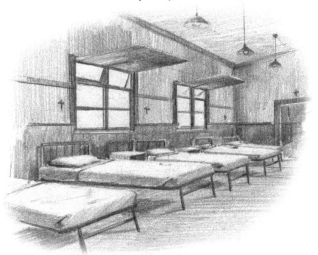

Joe arrived at the West End Military Hospital later that afternoon and addressed a pleasant looking young nurse at the front desk.

"I'm looking for Captain... Uh, Lieutenant Robert Browning" Joe asked.

"Browning..? Browning...? Let's see, Baker, Barker, Barrett, Blanchard, Bowen, Brophy, Bryant, Campbell." The nurse looked up. "Sorry, no Browning."

"What?" Joe was puzzled. "There must be some mistake. Lieutenant Robert L. Browning, 597th Fighter Squadron, 363rd Fighter Group, serial number 2134..." The nurse cut him off.

"I'm sorry Captain. I'll checked the list again, but like I told you, we don't have a Lt. Browning registered here."

Joe's confusion was beginning to turn into annoyance as he addressed the nurse once more.

"Well, do you mind if I have a look around anyway, nurse?"

"We have an awful lot of injured here Captain. Wounded come and go constantly." the nurse said skeptically, "But feel free to look around quietly if you'd like. Visiting hours are until 16:30."

Joe spent the rest of the afternoon searching every ward and bed, but found no trace of Rob or even anyone who knew of him. Each ward was filled with injured and dying men. Soldiers, Airmen, Sailors and support personnel all recovering from wounds received in the line of duty.

Burns, bullet wounds, shrapnel and missing limbs. Pain and suffering was everywhere, but there was no trace of his wingman.

On his way back to the front desk, Joe spotted a piece of paper crumpled up on the floor by a waste paper basket and feeling encouraged, bent down to retrieve it. Unfolding the sheet and hoping to find one of Rob's sketches, Joe's was crest-fallen to see the paper was blank.

At 16:29, Joe returned to the front desk where he was greeted by the same pleasant nurse he'd spoken to earlier.

"Did you find your friend?" she asked empathetically.

"No." Joe said looking defeated.

"Perhaps he was taken to one of these other hospitals?" The nurse offered handing him a list of names. Joe just nodded.

Placing his cap back on his head, Joe left the hospital wondering if he'd ever see the *Shadow* again.

For the remainder of his leave, Joe searched the London hospitals for his wingman without success. The large number of wounded coming and going daily made it very possible Rob had gotten lost in the military's shuffle. After three days of searching, Joe returned to Abington.

On May 7th, 1945, in Reims, France, representatives of the German High Command agreed to an unconditional surrender on all fronts, effective May 8th. Three weeks later, Captain Joseph M. Dyer of the 597th Fighter Group caught a troop ship out of Liverpool England bound for the United States. Joe was going home.

## My Shadow

The memories faded and back in his kitchen in Georgia, Joe read the last journal entry:

*June 26, 1945: Finally released from the hospital and caught the first train back to Abington only to find the base has been shut down and the 597th scattered to the four winds. It's hard to believe after so much time, it all just ends this way. No goodbyes, no thanks, no last words, just over. I received my travel orders today and will be shipping out myself next week. Have no idea what the future holds or what I'll do. Cushman, Taylor, Harrison, Wilson and the rest of Red Flight have already gone. Wonder if I'll ever see any of them again. And what about Georgia?*
*So much left unsaid - R.B.*

And one final entry appeared below in the same shaky hand as the address on the box.

*Georgia, too many years and not a day has gone by that I haven't thought of my Flight Leader. I have no clear recollection of our last mission other than waking in the hospital in a world changed forever. I hope that everything turned out all right for you, the way you always wanted. Sorry about my "butting in"... again, for one last time!*
*The Shadow knows?*
*R.B.*

# Robert Brun

## Epilogue
*Present Day*

Joe closed the journal and heard the front door open. His wife called out to him as she entered the house.

"In the kitchen." he called back, his voice cracking, the memories of his wingman fading back into the past.

Mrs. Maureen Dyer walked into the room and set her purse down on the kitchen table hanging her cane on the back of the chair. Looking over at Joe, she stopped, recognizing the expression she'd seen so many times over their last sixty-five years together.

Setting down the lighter, Joe stood up and hugged his wife tightly, the tears running down his face and onto her blouse. Mo hugged him back kissing him on the cheek.

"It's all right," she said gently stroking his hair. "It's all going to be all right."

**The End**

# My Shadow

## Glossary

*(German words in Italic)*

**4-F:** Draftees determined to be unfit for military duty for various medical or physical reasons.
**A.L.G.:** Air Landing Ground. Forward airfields hastily constructed often having primitive living conditions.
**Ack-ack:** Anti Aircraft fire.
**Balls to the Wall:** Pushing the throttle lever, topped with a round ball, all the way to maximum power.
**Barney Oldfield:** A flamboyant racecar driver popular in the 1920s.
***Baron Manfred Von Richthofen*:** 80 victory WW I German Ace known as the "Red Baron."
**Big Friends:** How fighter pilots referred to the bombers.
**Bingo:** The point where the fighter escort was require to turn back or risk running out of fuel.
**Bobbie:** A British Police Officer
*Boche:* A term used to refer to the Germans during WW I.
**Buy the Farm:** Reference to the $10,000 insurance payment received by families of soldiers killed in action. The money was often used to pay off the mortgage on "the farm."
**CO:** Commanding Officer.
**CAVU:** Ceiling And Visibility Unlimited. Clear skies to 10,000 feet +
**CCA:** Combat Command 'A.' Pilot assigned to a forward infantry unit responsible for calling in air strikes on targets.
**Cohiba:** High quality Cuban cigar.
**Collier's Weekly:** A popular illustrated publication during the 1940s.
**Contrails:** Visible vapor trails produced by aircraft during certain atmospheric conditions.
**CQ:** Charge of Quarters. The person in charge of waking pilots on their mission days.
**Dakota:** Name given to the C-47 transport plane. Also sometimes called the "Goony Bird."
**Dear John Letter:** A break-up letter from a sweetheart back home.
**Defensive Boxes:** A close flying formations used by USAAF bombers that enabled the bombers' gunners to give maximum protection to each other.
**Deuce and a Half:** A two and one half ton truck.
**Eagle Squadron:** The RAF Squadron made up of Americans that volunteered with the British prior to U.S. involvement in the war.
**Flak Happy:** Bomber's gunners who would shoot at any plane in range regardless of nationality.
**Gs or g-force:** The force of gravity felt on the body when climbing or turning (positive g) or diving (negative g).
**G.I. Bill:** Serviceman's Readjustment Act passed in 1944 to provide low cost, low interest loans for mortgages, businesses start-ups and collage tuition for returning G.I.s.
***Gruppe*:** German fighter group
**HP:** Hot pilot, a show off.
**HQ:** Head Quarters
**Hardstand:** A cement parking area usually surrounded by sandbags or a blast wall to protect parked aircraft from aerial attack.
**Heavies:** Four engine bombers.
**Immelman:** A flight maneuver named for WW-I Ace Max Immelman where by the plane pulls up and rolls in a climbing turn of 180 degrees.
**IP Initial Point:** The point ten minutes out from the target where the bombers begin their bomb run.
***Jadgeschwader*:** German fighter squadron.
***Jabos*:** Short for *Jadgbomber* or fighter-bombers.
**Jinxing:** Flying the plane in rapid turns and dives to throw off an opponents aim.
**Lamont Cranston:** A popular radio character whose alter ego was "the Shadow."
**Little Friends:** How bomber crews referred to fighter escorts.

# Robert Brun

**Lucky Strike Green Has Gone to War:** Slogan of the Lucky Strike cigarette company during WW II. Their Green brand label used copper for the ink, so they changed the label color for the war effort.
**Luftwaffe:** The German Air Force.
**Market Garden:** The Allied invasion of the Netherlands in September of 1944
**Marsden Matt:** Perforated interlocking steel mats used in rapid construction of forward airfields.
**Me and My Shadow:** Popular song of 1927 by Billy Rose and David Dreyer
**Meat Wagon:** The ambulance sent out to meet returning damaged aircraft.
**Milk Run:** An easy mission.
**Miss Grable:** 1940's actress Betty Grable know for having million dollar legs.
**MPs:** Military Police.
**Mutt & Jeff:** A popular comic strip by Bud Fisher about two friends, one short, one tall.
**Nissen Huts:** Corrugated half round steel structures used on air bases for housing, mess halls, etc.
**OC:** Officer's Club.
*Oswald Boelcke*: WW I German ace credited with developing a ten point guide for fighter pilots in combat.
**PDQ.** Pretty Dammed Quick!
**Peace For Our Time:** The comment made by then Prime Minister Neville Chamberlin after reaching an agreement with Adolf Hitler in 1939 to avoid armed conflict in Europe.
**Perspex:** An early type of Plexiglas.
**R&R:** Rest and recuperation.
**Ram Roding:** Flying bomber escort.
**Relief Tube**: A rubber tube, usually below the seat of a fighter used by the pilot to urinate.
*Schwarm:* Four German fighters flying line abreast. Also known as "Finger Four"
**Silk Elevator:** A parachute.
**Skivvies**: Underwear.
**SNAFU**: Situation Normal, All F**ked Up.
**SPAM:** Canned pork by-product produced in quantity by the Hormel$^{TM}$ Company to feed the troops during WW II.
**SPAM Can**: Slang term for a Mustang.
**Stars & Stripes**: Newspaper put out by the military for the troops during WW II.
**Terry & the Pirates:** A popular comic strip in the 1930 and 40s by Milton Caniff about the adventures of a pilot and his crew.
**Thousand Yard Stare**: The glazed look airmen took on after months in combat.
**Victory Roll**: Barrel rolling the plane, usually over the air base, once for each plane the pilot shot down and generally frowned upon by superiors as dangerous.
**V-Mail:** A weight saving process where letters are photographed, then shipped as negatives and reprinted for distribution.
**Vera Lynn:** A popular singer in England.
**WACO Gliders**: Towed wooded frame and fabric gliders used to silently transport troops and supplies often behind enemy line.
*Wehrmacht:* The German Army.
**Zippo**$^{TM}$: A popular type of cigarette lighter manufactured by the Zippo$^{TM}$ Company.

# My Shadow

## Aircraft Type

**Allied**
**A-20 Havoc**: Twin-engine medium bomber
**AT-6 Texan**: Advanced trainer flown before transitioning to fighters.
**B-17 Flying Fortress**: (Forts) Four engine heavy bomber built by Boeing.
**B-24 Liberator**: (Libs) Four engine heavy bomber built by Consolidated.
**B-26 Marauder**: (Widow Maker) Medium bomber built by Martin.
**C-47 Dakota**: Military version of the Douglas DC-3
**Hawker Hurricane**: (Hurribox) First British Monoplane fighter.
**P-38 Lightning**: Twin engine, twin tail fighter built by Lockheed.
**P-39 Aerocobra**: Single rear engine fighter built by Bell Aircraft.
**P-40 Warhawk**: Single engine fighter built by Curtis.
**P-47 Thunderbolt**: (T-Bolt or Jug) Single engine fighter built by Republic Aviation.
**P-51 Mustang**: Single engine fighter built by North American.
**Sunderland**: Four engine British flying boat built by Short/Sunderland.
**Spitfire**: (Spitter) Single engine British fighter built by Supermarine.

**German**
*(For earlier Messerschmitt fighters Bf is used instead of Me to indicate Bayerische Flugzeugwerke and later renamed Messerschmitt)*
**Bf-109**: The most produced single engine fighter of WW II designed by Willie Messerschmitt.
**Bf-110** *Zorstörer*/Destroyer: Twin engine, long-range bomber escort fighter designed by Willie Messerschmitt.
**FW-190** Focke-Wulf: Single engine fighter designed by Kurt Tank.
**Ju-88** Junkers: Twin-engine fast medium bomber.
**Me 262** *Schwable*/Swallow: First operational jet powered fighter designed by Willie Messerschmitt.
**V1** *Vergeltungswaffen* (reprisal weapon) *1:* Flying bomb powered by a pulsejet motor. Also called a "buzz bomb" or "Doodlebug."
**V2** *Vergeltungswaffen* (reprisal weapon) 2: Rocket fired to an altitude of 60+ miles and could exceed 2,000 mph.

**Robert Brun**

# My Shadow
## The Author

**Robert Brun:** Born in Baltimore, MD, Robert's interest in aviation started at a very early age. During the first 18 years of his life he built scale and flying models of just about *"anything with a propeller"*, but primarily planes of the 1940s

After attending the **Rhode Island School of Design** (R.I.S.D.) in the 1970's where he earned his B.F.A. in Illustration, Robert worked for various advertising agencies in Massachusetts and Rhode Island. In 1986 Robert founded the **Independent Pencil Company** where over the next 15 years he honed his artistic skills as a professional freelance illustrator. In 1998, after 18 years of commercial work, he made the transition to fine art specializing in oil and watercolor.

Moving to Newburyport, MA in 1985, where he currently lives and works, Robert continues to paint scenes depicting flight and the New England coast. His paintings can be seen at www.FighterArtist.com, and have been displayed at the Plum Island Aerodrome, The WASP WWII Air Museum in Sweetwater, TX and in private collections of pilots, sailors and art enthusiasts throughout the country. His paintings have appeared in the 2007 and 2009 American Society of Aviation Artists (ASAA, of which he is a member) juried show at BWI Airport in Baltimore, MD the 8th Air Force Museum in Savannah, GA and the Aviation History Museum in Kalamazoo, MI. His paintings *"Hurricane Scramble"* and *"Flight of the Tomahawk"* received Awards of Merit at the 2009-2010 CAE SimuFlite "Horizons of Flight" aviation art exhibits in Dallas, TX. He regularly works with the Collings Foundation in Stow, MA and is currently working with WW II veterans on a series of paintings from first hand accounts of aviation events. This is his first novel.

*"It is the excitement and freedom of flight along with the inherent beauty of the aircraft among the clouds that inspires my work. To be able to combine art with flying is truly the best of both worlds."*